Practical safety and reliability assessment

To my wife Sheila,
in grateful recognition of her encouragement
and assistance in the writing of this book

Practical safety and reliability assessment

K. C. Hignett

Consultant in Risk, Safety and Reliability
Northwich, UK
Chartered Engineer
Member – Institution of Electrical Engineers
Fellow of the Safety & Reliability Society

E & FN SPON
An Imprint of Chapman & Hall

London · Weinheim · New York · Tokyo · Melbourne · Madras

Published by
E & FN Spon, an imprint of Chapman & Hall, 2–6 Boundary Row,
London SE1 8HN, UK

Chapman & Hall, 2–6 Boundary Row, London SE1 8HN, UK

Chapman & Hall GmbH, Pappelallee 3, 69469 Weinheim, Germany

Chapman & Hall USA, 115 Fifth Avenue, New York, NY 10003, USA

Chapman & Hall Japan, ITP-Japan, Kyowa Building, 3F, 2-2-1 Hirakawacho, Chiyoda-ku, Tokyo 102, Japan

Chapman & Hall Australia, 102 Dodds Street, South Melbourne, Victoria 3205, Australia

Chapman & Hall India, R. Seshadri, 32 Second Main Road, CIT East, Madras 600 035, India

First edition 1996

© 1996 K. C. Hignett

Typeset in 10/12 Palatino by Type Study, Scarborough, North Yorkshire
Printed in Great Britain by TJ Press, Padstow, Cornwall

ISBN 0 419 21330 9

A catalogue record for this book is available from the British Library

Library of Congress Catalog Card Number: 96-67192

∞ Printed on permanent acid-free text paper, manufactured in accordance with ANSI/NISO Z39.48-1992 and ANSI/NISO Z39.48-1984 (Permanence of Paper).

Contents

Preface

Engineered systems generally are designed and operated to meet two primary objectives, namely to produce end products which will enhance society living standards and to realize adequate profit margins arising from the sales of the given products. In more simple terms, systems are designed to work successfully.

This book emphasizes to designers and operational managements why and how systems should also be designed to fail. System failures represent disorganization or increase in entropy and are always provided free of charge by nature itself in accordance with the second law of thermodynamics. In the absence of a failure design approach, overall system safety characteristics will be relatively unknown and will be accompanied by a higher risk of exceeding the boundaries of acceptability. With an engineered logical approach as explained within these pages, safety and reliability characteristics of systems can, with degrees of certainty, be directed towards known and acceptable criterial boundaries.

This practical guide has been written on experience drawn from over 40 years by the author in the chemical, oil and nuclear fuel processing industries. It presents essential working principles of safety and reliability engineering in forms which are compatible with general engineering requirements and practices. It is intended to inspire and encourage confidence on the part of plant engineers, operational managements and educational bodies to gain a closer understanding of its use and hence to promote its application at the basic engineering levels of systems design and operation. In practical terms it will enable essential foundations of the technology to be infused into everyday engineering activities.

Reliability, being statistical in nature, is not an exact science. It is founded mainly on experience of failures at the subsystem level and relies significantly on the application of a fault-tree-based logical approach which may be qualitative or quantitative. The contained subject matter deals with evaluation of systems reliability in terms of safety and related qualities which are capable of expression in engineering terms. Evaluations enable judicious comparisons with alternative designs in conceptual or existing plant systems and hence may be regarded as a powerful management tool which can lead to significant economies when directed towards criterial requirements of safety. As a final note of caution, it must, however, be stated that achievement of risk criteria through disciplined assessment and design can never guarantee that system failure leading to a defined hazard will not occur at any instant of time over some given period.

K. C. Hignett
Cuddington, Cheshire

Symbols

ROMAN

D	demand rate
D_p	percentage diversity
d	degrees of diversity
$F(t)$	exponential cumulative function
$f(t)$	failure probability density function
f	annual frequency of a given task
H	hazard rate
m	minimum number of subsystems for system success
n	number of subsystems
n_p	number of proof tests
n_s	number of elements in a cut set
N	number of faults in T years
P	failure probability
P_{av}	average failure probability
P_{CM}	common-mode failure probability
P_D	probability of a demand
P_H	human failure probability
P_I	independent-mode failure probability
P_{max}	maximum failure probability
P_0	system failure probability
P_R	revealed failure probability
$P(t)$	failure probability at time t
P_T	total failure probability
r	minimum number of subsystems for system failure
R	failure rate
$R(t)$	reliability up to time t_x
R_0	system failure rate

S	number of element pairs
S_a	number of active element pairs
S_p	number of element pairs in a parent cut set
t	time
t_∞	infinite time
T	total time in years
t_1	initial time
t_2	later time following initial time
t_x	future random instant of time
t_0	zero time
$Z(t)$	hazard function

GREEK

∞	infinity
β	ratio common-mode failure/total failure
β_{av}	average ratio common-mode failure/total failure
δ	finite increment $y = f(x)$
θ	failure rate
θ_{CM}	common-mode failure rate
θ_D	unrevealed equipment failure rate
θ_e	effective system failure rate
θ_H	human failure rate
θ_0	system failure rate
θ_R	revealed equipment failure rate
θ_r	single equipment revealed failure rate
μ	mean fractional dead time
μ_A	availability on demand
μ_{CM}	common-mode failure probability on demand
μ_D	unavailability on demand
μ_f	mean time between failures
μ_H	human failure probability on demand
μ_0	system unavailability
μ_R	revealed failure probability on demand
μ_r	system mean repair time
Σ	sum of all
τ	proof test interval
τ_m	inspection interval
τ_{op}	optimum proof test interval
τ_r	single equipment mean repair time

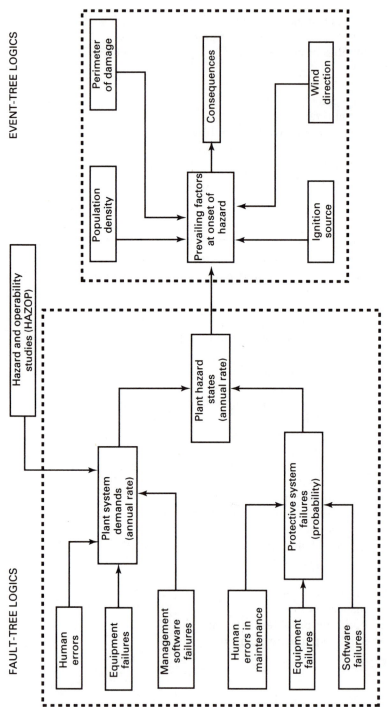

General logistics of process system safety engineering.

1

Terminologies in process safety engineering

1.1 PLANT SYSTEMS

1.1.1 Hazard states

Hazard states can be categorized as:

- **dangerous** process hazard states manifested by events which may lead to fatalities and/or injuries and/or plant system damage;
- **safe** process hazard states manifested by events which result solely in economic penalties due to loss of production consequent to enforced plant shutdown and when conditions implied by 'dangerous' are absent.

It should be recognized that those events which produce dangerous hazard states most often encompass monetary losses arising from loss of production, replacement cost of damaged plant and possible reparations to personnel.

1.1.2 Plant system demands

When considering safety, a demand is defined as a process system abnormality which, if allowed to proceed, would be expected to result in a dangerous process hazard state. Demands are normally assumed to be random or exponentially distributed with time.

1.1.3 Hazard and risk

The presence of any process abnormality which represents a dangerous or potentially dangerous state is known as a **hazard**. In high hazard scenarios, by way of example, it may refer to a detonation or perhaps a release of toxic materials. In lesser terms it may refer to, say an inadequate mixing of chemical constituents or build-up of dangerous by-products which may lead to the high hazard. Potential hazard states, particularly in chemical processing, are most often identified by carrying out hazard and operational studies known as HAZOPs.

Risk normally refers to the statistical annual frequency of the particular hazard state. For example, once per year would be very high risk, whereas once in 10 000 years would be considered a relatively low risk.

To illustrate the subtle difference between hazard and risk, consider a sharp-pointed pencil in the context of a hazard whereby it is capable of causing serious injury to a person's eye. When lying on a table, it poses an insignificant risk. However, if it should be taken up and pointed closely towards the person's eye then the risk of injury will be significantly greater.

1.1.4 Consequence

The presence of a process hazard state does not represent the ensuing consequences which may result from the said hazard when it becomes established. For example, taking a detonation as a hazard state, the range of possible outcomes could typically depend on such factors as proximity to personnel, containment etc. A single hazard state most often results in multiple outcomes which range from the insignificant to major accident scenarios. They are normally evaluated by the use of event-tree logics, characterized by a single input hazard event which propagates through multiple paths to outputs representing the range of possible consequences.

1.2 SAFETY SYSTEMS

A safety or safeguards system may be defined as a system of logic which is designed to await and recognize the onset of a specific process demand with the objective of preventing a dangerous process hazard. The commonly accepted design philosophy of a safety system is that it should be independent of the process system to which it affords the required safeguard. For purposes of clarification, independence implies that safety functions should not be derived from those measurement and control systems which provide operational process monitoring and control.

1.2.1 Reliability

Reliability may be defined as the probability or chance of a component, subsystem or composite system operating within its designed intent within the boundaries of required function and environment for a given stated period of time.

Reliability may be expressed in terms of **success** or **failure**, i.e.

$$\text{Success probability} = \text{Reliability}$$

$$\text{Failure probability} = 1 - \text{Reliability}$$

Hence success and failure are mutual complements.

1.2.2 Capability

Capability implies that the system or subsystem will function within defined boundaries of operational and environmental requirements. As a precursor to any reliability assessment, it is mandatory that the engineer or assessor must be satisfied that capability will be present in the system or subsystem design and installation when free from failures. By way of example, and for purposes of clarification, a temperature sensor located in a non-representative position in a process stream would not have the capability to provide the safeguard requirement irrespective of its inherent reliability.

1.2.3 Safety logic

- **Redundancy** refers to a subsystem which consists of more than one protective channel arranged such that failure of any single channel will not negate the desired safety function of the subsystem group.
- **Diversity** is a term applied to redundant safety subsystems and implies that the separate channels of that system are not identical to each other but nevertheless do each provide the stated common protective function.
- **Voting logic** is a term which describes the way in which the outputs from the elements of a redundant system are combined in order to initiate a desired safety action. Voting is used to minimize spurious or nuisance safety actions whilst at the same time achieving the desired reliability of safeguard function.

1.2.4 Safety system failure modes

(a) Dangerous modes

In safety systems, the failures of prime interest are those which would prevent the system from initiating a safe plant shutdown should a dangerous process demand arise. Such failures are termed dangerous.

(b) Safe modes

Faults which arise in safety systems which initiate plant shutdowns even though process demands are absent are defined as safe or spurious failures.

1.2.5 Fault classifications

(a) Unrevealed faults

These are the failures of paramount interest in safety systems and are therefore termed **dangerous unrevealed faults**. When present the safety system would exhibit normal characteristics of healthy working but would not respond to a process demand. Such failures are undetectable except at the time of a **proof test**. Generally the unrevealed fault must always be related to the specific mode of system operational requirement. A diesel generator failing to start on a process system demand is an unrevealed fault mode, but would normally be regarded as **safe** since the process would be expected to shut down safely as a consequence of total power supply failure. The same mode of failure would be dangerous if related to a life support system.

(b) Revealed faults

The presence of revealed or spurious faults in safety systems is usually regarded as safe since they initiate process shutdowns with no plant demands or dangerous abnormalities being present. In the plant system itself a pump system failure or automatic process valve spurious closure could also be expected to cause the process to shut down to a safe state.

1.2.6 Failure dependences

(a) Independent failures

Certain random fault modes in both the revealed and unrevealed categories are classified as being independent in nature. Failure independence means that all relevant faults of interest in the engineered subsystems are assumed to have no underlying common cause and are free of mutual interactions in terms of failure propagation. It is an unfortunate reality that data bank sources are not able to adequately identify purely independent failure elements in collected field data. In the practical world of engineering, it is the practice to assume failure data to be initially independent.

(b) Dependent failures

These failures are also assumed to be randomly distributed in time, and as the name implies they have dependences on common causes. For this reason they are referred to as either **dependent, common-cause** or **common-mode** failures, the last designation being the most often used in general engineering. In order to appreciate the nature of their occurrence it is most convenient to consider a number of identical subsystems arranged in a redundant system configuration. Each of the system elements will have an assigned identical but independent failure rate derived from some data source. However, there will be in addition, and contained within the data, a failure mode element due to some common cause which, if present, would bring about failure of all the subsystems in the redundant group. Data on this type of failure are very difficult and costly to acquire and are virtually unobtainable for general engineering purposes from established data bank sources.

1.2.7 Failure rate

Failure rate is defined as the average number of failures per unit time for a given item or subsystem based on environment and operating experience of a sample population of such items over its total integrated time of operation.

1.2.8 Failure distribution

Failure distributions describe the statistical manner in which failure rates are related to time. Some of the more well known examples of time-dependent distributions are those of the exponential, normal, log normal and Weibull types. Most engineered components and subsystems exhibit time dependences, but in general such data are not easy to acquire due to the difficulties and high cost of acquisition.

Systems reliability engineering practice assumes an exponential distribution of failure rate which is constant with time and randomly distributed over the useful life phase of the equipment in use. At first sight this assumption may be seen to be invalid since it is known that many components and subsystems clearly do not exhibit constancy of failure rate over the useful life phase. A typical example would be a type of centrifugal pump used in a sewage treatment works which transfers raw sewage influent containing grit to the primary settlement stage. In such a pumping system of, say two or more pumps, it is known that each pump unit is operating in a wearout phase distribution and hence is time dependent to failure, i.e. it exhibits an increasing failure rate with time. The individual pumps which constitute the subsystems of the pumping system will, however, reach unacceptable failure states in terms of operational capability singularly at different times. The incidence of pump failures will therefore undergo randomization with time and hence approximate to the exponential distribution, i.e. they will exhibit an average failure rate which is influenced by the duty environment. It is fortuitous that randomization of failures in components and subsystems approximates to the exponential failure distribution which is less rigorous in mathematical terms.

1.2.9 The bathtub curve

Most engineered components, subsystems and systems exhibit a failure rate characteristic with time known as the 'bathtub curve' shown in Figure 1.1, and so called because of its characteristic shape. It can be divided into three phases as follows.

- **Early life** or **infant mortality phase** (A): the failure rate is time dependent and initially high, but falls quite rapidly. Failures in this phase are predominantly due to manufacturing and/or installation/commissioning faults, hence the need for good quality control, adequate soak testing and pre-commission testing before being put into service.

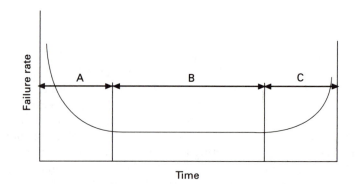

Figure 1.1 The bathtub curve.

- **Useful life phase** (B): the failure rate is almost constant against time and therefore has conformity with the exponential failure distribution with actual failures being due to undefined random occurrence of faults. In reliability engineering, it is common practice to operate equipment in the useful life phase.
- **Wearout phase** (C): the failure rate is increasing with time and is due to a known reason, namely wearout. Components and sub-systems in safety-orientated systems generally need to be replaced before they enter this phase.

1.2.10 Proof test interval

This is normally expressed in terms of years and refers to a periodic function test for evidence of continued system capability. Proof tests are applied to protective systems in order to demonstrate the presence or absence of those unrevealed dangerous failures which would otherwise degrade the capability of a given safety system to respond adequately to a random demand from the process system.

Ideally the test should demonstrate both accuracy and dynamic response within the designed limits of acceptable capability. It also assumes that at the end of each test the system is fully restored to its initial fault-free state. In the exponentially distributed model, the proof test is significant to reliability performance and is essential to maintaining the safety equipment within its useful life phase.

1.2.11 Mean fractional dead time (FDT)

Mean fractional dead time is defined as the failure probability of a protective system on demand. In the exponentially distributed model, system failure probability varies from zero at the beginning of the proof test interval to a maximum when the interval expires. Since all process demands are considered to be random (exponentially distributed) it follows that a demand is equally probable at any instant over the proof test interval. Hence the failure probability of a protective system is pertinent to its average value or FDT over the interval. It also represents the **average statistical time** in which the protective system is in a failed state, expressed as a fraction of the proof test interval.

2

Derivation of basic formulae

2.1 INTRODUCTION

The mathematical approach which follows seeks to establish practical basic formulae which are used in safety and reliability assessments.

2.2 FAILURE PREDICTION

Consider a safety device or system which is required to provide a desired safeguard action when a random process demand is applied to its input terminals. Initially at zero time following testing and possible repair, the equipment will be seen to be operating in a fault-free state. The salient point of interest is concerned with the chance, or probability, that the safety device may be unable to respond to a future demand due to the onset of some earlier undefined random fault.

A safety device which fails at some future unknown time t_x on the basis that it was free of faults at zero time t_0 requires the following two conditions to be satisfied.

- It will survive up to future time t_x which is a measure of its **survival probability** and hence **reliability**.
- It will fail at future time t_x which is a measure of its **instantaneous failure rate** at that time.

2.3 RELIABILITY FUNCTION

Reliability is the chance or probability of a device surviving up to a given time t_x. In order to express reliability, consider a large population of the devices such that they were all working at zero time t_0. Hence in satisfying the first requirement of failure prediction:

$$\text{Reliability } R(t) \text{ up to time } t_x = \frac{\text{Survivors at } t_x}{\text{Population at } t_0}$$

$R(t)$ is known as the **reliability function**.

2.4 HAZARD FUNCTION

The instantaneous failure rate of a given device at some future time t_x which satisfies the second requirement of failure prediction is expressed as

$$\text{Failure rate } Z(t) \text{ at time } t_x = \frac{\text{Failures/hour at } t_x}{\text{Survivors at } t_x}$$

$Z(t)$ is known as the **hazard function**.

2.5 FAILURE PROBABILITY DENSITY FUNCTION

This is an expression of failure rate for a device at any future time t_x on the basis that it was working and free of faults at zero time t_0. It is designated as $f(t)$ in reliability engineering and is the product of $R(t)$ and $Z(t)$:

$$R(t)Z(t) = \frac{\text{Survivors at } t_x}{\text{Population at } t_0} \times \frac{\text{Failures/hour at } t_x}{\text{Survivors at } t_x}$$

then

$$R(t)Z(t) = \frac{\text{Failures/hour at } t_x}{\text{Population at } t_0} = f(t) \tag{2.1}$$

Therefore

$$f(t) = R(t)Z(t) \tag{2.2}$$

The density function $f(t)$ has the dimension of rate, not probability as the name may imply. This will become apparent when the function in its exponential form has been derived.

$f(t)$ is known as the failure probability density function since if it be summed up to infinite time t_∞ it will reach a limiting value of unity:

$$\sum_{t=0}^{t=\infty} \frac{\text{No. failing/hour at each increment of time } t_x}{\text{Population at } t_0} = 1.0$$

For the exponential failure distribution, the failure rate θ is independent of time and hence $Z(t)$ is constant. Let $Z(t) = \theta$, then

$$R(t)\theta = f(t) \tag{2.3}$$

2.6 THE EXPONENTIAL RELIABILITY FUNCTION $R(t)$

The function $R(t)$ will be shown to be exponential when the failure rate is constant. At zero time t_0 there are no failures, hence it has a value of unity which decreases exponentially to zero at infinite time t_∞. This exponential decay can be conveniently demonstrated by reference to a hypothetical population of, say 1000 samples which is assumed to have an average failure rate of, say 0.1 per annum. Table 2.1 shows values of $R(t)$ for a number of equal time intervals. Plotting the values of t and $R(t)$ from this table produces the reliability characteristic shown in Figure 2.1. It can be seen that $R(t)$ decreases exponentially in value as t increases such that $R(t) = 0$ at infinite time t_∞.

Table 2.1 Reliability function over equal time intervals

Time	Number of failures	Survivors	$R(t)$
t_0	0	1000	1.0
t_1	100	900	0.9
t_2	90	810	0.81
t_3	81	729	0.73

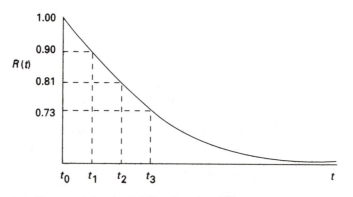

Figure 2.1 Characteristic of reliability function $R(t)$.

2.6.1 Derivation of the exponential law for $R(t)$

In order to establish the law for $R(t)$ the curve in Figure 2.1 is reproduced in Figure 2.2, from which an expression for gradient and hence
$R(t)$ will be derived. Referring to Figure 2.2, the average gradient of
the curve between the two points shown is:

$$\frac{\delta R(t)}{\delta t} = \frac{R(t_2) - R(t_1)}{t_2 - t_1}$$

$$= \left(\frac{\text{Survivors at } t_2 - \text{Survivors at } t_1}{\text{Population at } t_0}\right) \bigg/ \delta t, \quad \delta t = t_2 - t_1$$

$$= -\left(\frac{\text{No. of failures over interval } \delta t}{\text{Population at } t_0}\right) \bigg/ \delta t$$

$$= -\frac{\text{Average failures per hour over interval } \delta t}{\text{Population at } t_0}$$

Let $\delta t \to 0$, then

$$\frac{dR(t)}{dt} = -\frac{\text{Failures per hour at } t_1}{\text{Population at } t_0} = -f(t)$$

hence

$$f(t) = -\frac{dR(t)}{dt}$$

from which equation 2.3 can now be rewritten as

$$R(t)\theta = -\frac{dR(t)}{dt}$$

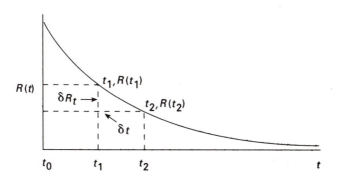

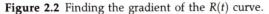

Figure 2.2 Finding the gradient of the $R(t)$ curve.

Transposing terms,

$$\frac{dR(t)}{R(t)} = -\theta\,dt$$

Integrating both sides,

$$\int_0^t \frac{dR(t)}{R(t)} = -\int_0^t \theta\,dt$$

i.e.

$$\log_e R(t) = -\theta t$$

hence

$$R(t) = e^{-\theta t} = \text{Survival probability} \tag{2.4}$$

from which

$$\text{Failure probability } F(t) = 1 - e^{-\theta t} \tag{2.5}$$

2.7 THE EXPONENTIAL DENSITY FUNCTION $f(t)$

It has already been shown that the exponential failure probability density function $f(t) = R(t)\theta$. Referring to equation 2.4

$$R(t) = e^{-\theta t}$$

therefore

$$\frac{dR(t)}{dt} = -\theta e^{-\theta t}$$

hence

$$-\frac{dR(t)}{dt} = \theta e^{-\theta t}$$

i.e.

$$f(t) = \theta e^{-\theta t} \tag{2.6}$$

The exponential characteristic for $f(t)$ is shown in Figure 2.3. It can be seen that $f(t)$ is a rate function since if $t = 0$ then $f(t) = \theta$, which is a failure rate.

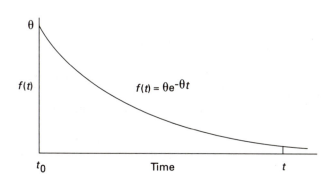

Figure 2.3 Characteristic of failure probability density function $f(t)$.

2.8 THE EXPONENTIAL CUMULATIVE FUNCTION $F(t)$

Failure probability $F(t)$, which has previously been shown to be the complement of $R(t)$, is represented by the area under the curve $f(t)$ shown in Figure 2.3. It is also known as the **cumulative function**. Hence failure probability up to time t is given by

$$F(t) = \int_0^t f(t)\,\mathrm{d}t = \int_0^t \theta e^{-\theta t}\,\mathrm{d}t$$

$$= \theta \left[\frac{e^{-\theta t}}{-\theta}\right]_0^t$$

$$= \theta \left[\frac{e^{-\theta t}}{-\theta} - \left(\frac{-1}{\theta}\right)\right]$$

$$= \theta \left(\frac{1}{\theta} - \frac{e^{-\theta t}}{\theta}\right)$$

hence

$$F(t) = 1 - e^{-\theta t} \tag{2.7}$$

The cumulative function $F(t)$ expresses failure probability in relation to time t. It is most commonly symbolized by $P(t)$ in preference to $F(t)$, i.e. in general terms

$$P(t) = 1 - e^{-\theta t}$$

where θ is the failure rate when exponentially distributed, and t is the time interval of interest.

If $\theta t < 0.1$ then $P(t) = \theta t$.

2.9 SUMMARY AND FINAL COMMENTS

The basic formulae derived in this chapter are summarized graphically in Figure 2.4.

Note that $f(t) = R(t)Z(t)$ is true for most distributions. However, engineered systems mostly conform to the exponential failure distribution, and its use is therefore valid for most practical quantitative assessment purposes. Distributions which are characterized by time-dependent failure rates involve rigorous mathematical treatments, but are found in practice to randomize with time into the exponential form; hence its wide use in quantitative reliability assessments.

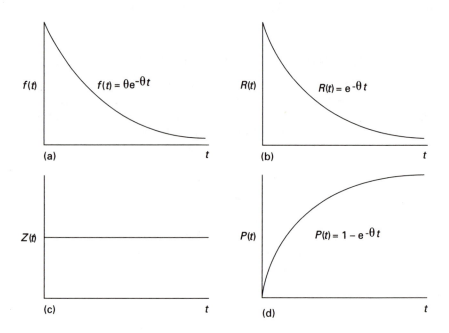

Figure 2.4 Summary of basic formulae for (a) density, (b) reliability, (c) hazard and (d) cumulative functions.

3

Unavailability, safety and changes of state

3.1 FAILURE MODES AND PLANT UNAVAILABILITY

In order to distinguish between plant unavailability and any relationship it may have to safety, it is necessary to consider the modes of failure which are met in assessment studies. In terms of plant unavailability, failures which are classified as safe do not usually involve safety but rather the loss of some plant or equipment function which would result in unscheduled or spurious plant shutdown and hence loss of production. However, where safety is the prime consideration, failures of interest are classified as dangerous since they are capable of resulting in injuries to operating personnel and/or plant damage. Both safe and dangerous modes of failure will of course lead to plant unavailability.

3.2 FAILURE RATE COMPONENTS

Failure rates are generally expressed in total value form and therefore it is always necessary to consider the mode of particular interest and thence to extract the relevant component from the overall value. In simple terms, by way of example, a pressure switch will have three modes of failure, each with a particular failure rate such that their sum would represent the overall failure rate of the switch. For clarification, each failure mode will be discussed in order to indicate how each could be viewed in an assessment in either safety or plant unavailability terms. The switch is considered in a safety function role.

3.2.1 Failure producing the effect of a maximum input signal

- If plant shutdown is required when the process generates a maximum input signal then the failure is rated as a **safe revealed** fault since it shuts the process down even though a demand is absent. It is therefore related only to plant **unavailability**.
- If shutdown is required when the process generates a minimum input signal then the failure is rated as a **dangerous unrevealed** fault since it could not shut the process down should a demand occur. It is therefore related to **safety** and **potential plant unavailability** if damage should occur.

3.2.2 Failure producing the effect of a minimum input signal

- If shutdown is required when the process generates a minimum input signal then the failure is rated as a **safe revealed** fault since it shuts the process down even though a demand is absent. It is therefore related only to plant **unavailability**.
- If shutdown is required when the process generates a maximum input signal then the failure is rated as a **dangerous unrevealed** fault since it could not shut the process down should a demand occur. It is therefore related to **safety** and **potential plant unavailability** if damage should occur.

3.2.3 Failure producing the effect of a normal input signal

- In **safety** terms this is a **dangerous unrevealed** fault since it prevents plant shutdown irrespective of whether shutdown is required on the evidence of a high or low input signal. It is therefore related to **safety** and **potential plant unavailability** if damage to the plant should result.
- In **plant unavailability** terms the failure has no relevance since it does not result in an unscheduled plant shutdown.

3.2.4 Symbolic representation of failure modes

As a further aid to clarity let the three failure modes in sections 3.2.1, 3.2.2 and 3.2.3 above be designated with failure rate symbols θ_1, θ_2 and θ_3 respectively. The overall switch failure rate is given as θ_0. Hence

$$\theta_0 = \theta_1 + \theta_2 + \theta_3$$

- Consider the safety mode of operation and evaluate the relevant dangerous failure rates for both the high and low trip modes, noting that 'trip' refers to automatic shutdown action.

(a) High trip:

> Unrevealed dangerous failure rate $= \theta_2 + \theta_3$

(b) Low trip:

> Unrevealed dangerous failure rate $= \theta_1 + \theta_3$

- Consider the plant unavailability mode of operation.
  (a) High trip:

> Revealed safe failure rate $= \theta_1$

(b) Low trip:

> Revealed safe failure rate $= \theta_2$

3.3 CHANGES OF STATE

Equipment can operate in any one of three states:

- irreversible;
- partly reversible;
- reversible.

3.3.1 Irreversible state

This state implies a system whereby repair is not possible in the event of a failure, hence the system cannot be tested with a view to repair. A typical system would be a satellite or deep space probe which cannot be retrieved for repair and which does not have an adequate programmed on-board repair facility. Hence for this state the failure probability over any given time of interest t is given by

$$P = 1 - e^{-\theta t}$$

where P is the failure probability over time t; θ is the failure rate in the mode of interest.

3.3.2 Partly reversible state

This is the state relevant to safety modes of operation in which the faults of interest are those of the unrevealed type which can only be found by carrying out proof tests. The state is said to be partly reversible since repairs are not possible between successive proof tests. When the test takes place the system is seen to be either fully operational or in the failed state or exhibiting conditions leading to a failed state. At the end of the proof test procedure the system is assumed to be fully restored to its original working state. Proof testing time and repair time are usually insignificant compared to the proof test interval and therefore can normally be ignored.

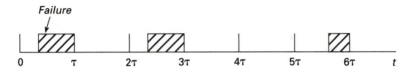

Figure 3.1 The partly reversible state.

(a) Working formulae in the partly reversible state

Figure 3.1 depicts the partly reversible state with proof tests carried out at intervals of time τ. A random failure persists until the end of the proof test interval.

There are two points of interest:

$$\text{Availability } \mu_A = \frac{\text{Time in the working state}}{\text{Total time}}$$

$$\text{Unavailability } \mu_D = \frac{\text{Time in the failed state}}{\text{Total time}}$$

It is now necessary to consider how to express time in the working state or time in the failed state when in fact there is no knowledge of the actual time of failure in the proof test interval. To meet this requirement it is necessary to resort to the idea of statistical time, i.e. to base it on the probable duration of operation or failure, the pertinent fault mode being that of the unrevealed or dangerous with the assumption of an exponential distribution of failure. Now consider the basic cumulative function $P = 1 - e^{-\theta t}$ which is illustrated in Figure 3.2. This represents the cumulative failure probability over the test interval τ.

Average failure probability over the interval τ is given as

$$P_{\text{av}} = \frac{1}{\tau} \int_0^\tau (1 - e^{-\theta t})\, dt$$

If $\theta t \ll 1.0$, then

$$P_{\text{av}} = \frac{1}{\tau} \int_0^\tau \theta t\, dt = \frac{\theta \tau}{2}$$

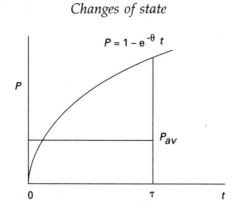

$$P = 1 - e^{-\theta} t$$

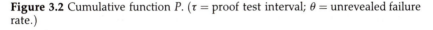

Figure 3.2 Cumulative function P. (τ = proof test interval; θ = unrevealed failure rate.)

The expression $\theta\tau/2$ is known as the **mean fractional dead time** (FDT) and is defined as the failure probability on demand or unavailability at any instant of time within the proof test interval for a single system. It also represents the statistical fraction of any given time interval over which failure will be present.

Unavailability μ_D already generally defined above may now be expressed in more precise terms as follows:

$$\mu_D = \frac{\text{Probable time in the failed state}}{\text{Proof test interval}}$$

$$= \frac{(\theta t/2)\tau}{\tau}$$

$$= \frac{\theta t}{2}$$

Hence:

- μ_D is the unavailability of the partly reversible state;
- μ_A is the availability of the partly reversible state.

Hence

$$\mu_D = 1 - \mu_A$$

$$\mu_A = 1 - \mu_D$$

Failure probabilities on demand applicable to the partly reversible state can now be conveniently summarized as follows:

- **Mean fractional dead time** μ_D of a system or subsystem is the probability of the system being in a failed state when a random

demand is made. It expresses the unavailability of the partly reversible state.

- **Mean fractional available time** μ_A of a system or subsystem is the
  probability of the system being in a normal operating state when a
  random demand is made. It expresses the availability of the partly
  reversible state.

3.3.3 Reversible state

This state has a significant relevance in safeguards engineering and will
therefore be dealt with in sufficient detail in this section. Although the
state is also pertinent to process plant equipments the emphasis will
be on the safeguards aspect in the discussions which follow.

Spurious operations of safeguards systems produce false alarms and/
or unscheduled plant shutdowns in the absence of plant demands and
therefore give rise to economic penalties by way of plant production
losses or lesser significant nuisance events.

In designing safeguards systems the principal requirement is to meet
the reliability criteria necessary to satisfy the incidence of plant
demands. However, in meeting such criteria it is also necessary to
ensure that the occurrence of spuriously designated safe operations are
of an acceptable frequency.

Figure 3.3 depicts the reversible state which may be defined as one in
which a failure condition is immediately communicated. Repair or
renewal can commence from the time of failure and, since such failures
are revealed, proof testing is not carried out. This is the state of prime
interest in availability modelling and raises two points of interest:

$$\text{Availability } \mu_A = \frac{\text{Time in the working state}}{\text{Total time}}$$

$$\text{Unavailability } \mu_D = \frac{\text{Time in the failed state}}{\text{Total time}}$$

In this state repair times are significant since the plant is shut down
or without protection whilst the repair is being carried out.

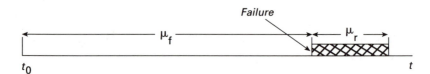

Figure 3.3 The reversible state.

Revealed faults in the safety engineering mode to which the reversible state applies are as follows.

- **Revealed fail to danger faults:** these imply that a fault has occurred which prevents operation of the safety system should a plant demand arise. The equipment fault is apparent and repair or replacement can be carried out before the next scheduled proof test. The apparent mean time between faults will be the sum of the mean time between revealed failures and the mean repair time.
- **Revealed fail safe (spurious) faults:** these are revealed since they produce a plant shutdown or false alarm, or an approach to spurious shutdown. The apparent mean time between faults will be the sum of the mean time between spurious faults and the mean repair time.

Referring back to Figure 3.3, expressions will now be derived for availability μ_A and unavailability μ_D. Availability is dependent on two factors:

- mean time between failures (μ_f);
- mean time to repair (μ_r).

Now

$$\mu_A = \frac{\text{Time in the working state}}{\text{Total time}}$$

$$= \frac{\text{Mean time between failures}}{\text{Mean time between failures} + \text{Mean repair time}}$$

Hence

$$\mu_A = \frac{\mu_f}{\mu_f + \mu_r}$$

from which

$$\mu_D = 1 - \mu_A$$

$$= \frac{\mu_r}{\mu_f + \mu_r}$$

$$= \frac{\mu_r}{\mu_f}$$

if μ_r is insignificant compared to μ_f.

(a) Working formulae for the reversible state

Mean fractional dead time and availability

Let θ_r be the revealed fault rate (single equipment), τ_r the mean repair time (single equipment) and T the total time in years. The mean time between failures is $1/\theta_r$ (reciprocal of failure rate is time).

When a failure occurs the equipment remains out of service for a further period equal to the mean repair time τ_r. Therefore

$$\text{Apparent mean time between failures} = \frac{1}{\theta_r} + \tau_r$$

$$= \frac{1 + \theta_r \tau_r}{\theta_r}$$

$$\text{Apparent revealed failure rate } \theta_e = \frac{\theta_r}{1 + \theta_r \tau_r}$$

Let N be the number of faults in total time T years. Then

$$N = \frac{\theta_r T}{1 + \theta_r \tau_r} \qquad \text{(i.e. faults/annum} \times \text{no. of years)}$$

Now

$$\text{Total dead time} = \text{Total accumulated repair times}$$

Let τ_f be the mean total dead time. Then

$$\tau_f = N\tau_r = \frac{\theta_r T}{1 + \theta_r \tau_r} \times \tau_r$$

$$\text{Mean fractional dead time } \mu_D = \frac{\text{Mean total dead time}}{\text{Total time}} = \frac{\tau_f}{T}$$

Therefore

$$\mu_D = \frac{\theta_r \tau_r}{1 + \theta_r \tau_r} = \frac{\mu_r}{\mu_f + \mu_r}$$

If $\theta_r \tau_r \ll 1.0$ then $\mu_D = \theta_r \tau_r$.

$$\text{Availability } \mu_A = 1 - \mu_D$$

$$= 1 - \frac{\theta_r \tau_r}{1 + \theta_r \tau_r}$$

$$= \frac{1}{1 + \theta_r \tau_r} = \frac{\mu_f}{\mu_f + \mu_r}$$

Effective system failure rates
In availability modelling of the reversible state it is frequently required to know how often per year that a plant is shut down due to spurious safe operations of the protective system. Such faults arise in items of plant equipment and are always capable of producing unscheduled shutdowns whether the logic is of simple or redundant form with varying degrees of majority voting. Non-redundant safety logics rarely provide adequate levels of inherent reliability, hence it is the practice to employ redundancy in order to meet higher safety requirements. Majority votings in safety logics provide higher levels of protection with added benefits of minimal spurious operations of the safety system. Hence for availability modelling of safety systems where effective revealed failure rates are used the following methods are illustrated for determining overall system spurious trip rates in series redundancy and two-out-of-three majority voting logics.

Single system It has previously been shown that the effective spurious failure rate θ_e for a single equipment is given by

$$\theta_e = \frac{\theta_r}{1 + \theta_r \tau_r}$$

If $\theta_r \tau_r \ll 1.0$ then $\theta_e = \theta_r$.

n series system

$$\theta_e = \frac{n\theta_r}{1 + \theta_r \tau_r}$$

If $\theta_r \tau_r \ll 1.0$ then $\theta_e = n\theta_r$.

Majority voting – two-out-of-three system Consider three protective channels A, B and C, each with an effective spurious failure rate of θ_A, θ_B and θ_C respectively and determine the effective system spurious trip rate θ_e. The mean repair time of any of the three channels is given as τ_r.

The majority vote 'two out of three' specifies that any two of the three channels must trip before an automatic plant system shutdown can take place.

In general terms the mean fractional dead time (FDT) of each channel is given as

$$\mu_D = \theta\tau_r \quad \text{(assuming that } \theta\tau_r \ll 1\text{)}$$

- Condition 1: Let A fail first followed by **either** B or C.
  Unavailability or FDT of $A = \theta_A \tau_r$

Rate of trip due to B failing after $A = \theta_B(\theta_A \tau_r)$

Rate of trip due to C failing after $A = \theta_C(\theta_A \tau_r)$

- Condition 2: Let B fail first followed by **either** A or C. Unavailability or FDT of $B = \theta_B \tau_r$

  Rate of trip due to A failing after $B = \theta_A(\theta_B \tau_r)$

  Rate of trip due to C failing after $B = \theta_C(\theta_B \tau_r)$

- Condition 3: Let C fail first followed by **either** A or B. Unavailability or FDT of $C = \theta_C \tau_r$

  Rate of trip due to A failing after $C = \theta_A(\theta_C \tau_r)$

  Rate of trip due to B failing after $C = \theta_B(\theta_C \tau_r)$

The effective rate of trip θ_e of two out of three will be the sum of rates derived from the above three conditions, i.e.

$$\theta_e = 2\tau_r(\theta_A\theta_B + \theta_A\theta_C + \theta_B\theta_C)$$

If $\theta_A = \theta_B = \theta_C = \theta$, then

$$\theta_e = 6\theta^2 \tau_r$$

General equation for safety system spurious trip rate
Let

θ_e = system effective spurious trip rate

θ = single subsystem apparent spurious trip rate

τ_r = mean repair time of a single subsystem

n = total number of subsystems

m = minimum number of subsystems required for system success

r = minimum number of subsystems required to give system failure

Note that

$$r = n - m + 1.$$

The equation is given as follows:

$$\theta_e = \left|\begin{array}{c} n \\ r \end{array}\right| r\theta(\theta\tau_r)^{m-1}$$

For example, for a two-out-of-three system, $r = 2$, $n = 3$, $m = 2$:

$$\theta_e = \frac{3 \times 2}{1 \times 2} \times 2\theta(\theta\tau_r)$$

$$= 6\theta^2 \tau_r$$

Also for a two-out-of-four system, $r = 3$, $n = 4$, $m = 2$:

$$\theta_e = \frac{4 \times 3 \times 2}{1 \times 2 \times 3} \times 3\theta(\theta\tau_r)$$

$$= 12\theta^2\tau_r$$

Summary of formulae for the reversible state

- Mean fractional dead time:

$$\mu_D = \frac{\mu_r}{\mu_r + \mu_f} = \frac{\mu_r}{\mu_f}$$

  if μ_r is insignificant compared to μ_f
- Mean fractional available time:

$$\mu_A = \frac{\mu_f}{\mu_f + \mu_r}$$

- Effective single system spurious failure rate:

$$\theta_e = \frac{\theta_r}{1 + \theta_r\tau_r} = \theta_r \qquad (\text{if } \theta_r\tau_r \ll 1.0)$$

- Effective spurious failure rate for n series system:

$$\theta_e = \frac{n\theta_r}{1 + \theta_r\tau_r} = n\theta_r \qquad (\text{if } \theta_r\tau_r \ll 1.0)$$

(b) Downtime and failure rate

It has been shown that availability or unavailability is dependent on two factors:

- safe or revealed failure rate;
- mean repair time or downtime.

Downtime is more commonly referred to as repair time and is an important factor in achieving desired levels of systems availability. In an ideal hypothetical situation where repair could be carried out instantaneously without interruption of service a given system availability would be 100%. However, in the practical world, failure necessitates repair and whilst this is being carried out the system is out of service.

   The annual incidence of repair is dependable on the annual frequency of failures in that every failure event demands a maintenance operation. Hence the failure rate of a system is seen to be a significant contributor to overall system availability. In designing systems, equipment should be selected with the most practical favourable failure rates. The achievement of optimum failure rate is not usually under the direct control of the user apart from choice of type, but is more largely dependent on outside manufacturing influences.

Repair times are, on the other hand, significantly under the control of the user and hence, because of potential flexibility of in-house maintenance, unfavourable failure rates can be offset by judicious planning of repair regimes. Overall downtime consists of a number of elements of which the more significant ones are:

- time to reveal from onset of fault;
- reporting procedures;
- administration time;
- availability of spares;
- availability of maintenance;
- removal from plant stream;
- repair time in workshop or plant;
- re-installation time;
- commissioning and testing.

4

Proof testing

4.1 INTRODUCTION

The failure probability of a protective system is dependent on both proof testing and failure rate. The former, for a given failure rate, has the primary function of ensuring that the established reliability of the system is maintained within the useful life phase as indicated in phase B of the bathtub curve (Figure 1.1).

Failure rate is an inherent quality in safety engineering terms of equipment reliability in its useful life phase and is therefore deemed to be independent of time when the exponential distribution of failure is assumed. This implies that any replacement of components as a result of maintenance or testing will be subject to a satisfactory 'soak test' or 'burn-in period' in order to eliminate manufacturing or commissioning faults, i.e. time-dependent early life failures.

Safety systems are periodically proof tested in order to reveal any dangerous unrevealed faults which may have occurred since the previous test and which would degrade or negate the required protective capability of the system. A dangerous unrevealed fault is classified as a failure which is not readily observable and can therefore only be detected at the time of a proof test. The test requirement is to demonstrate continued capability in terms of accuracy and dynamic response necessary to meet the most pessimistic process dynamic condition.

4.2 PROOF-TESTING PHILOSOPHIES

The ideal proof test would be one whereby the process would be intentionally taken into the worst credible dynamic demand state up to, but not exceeding, the desired automatic safety datum limit and the ensuing action by the safeguards system verified. Clearly this would not be practicable for both technical and plant safety reasons.

In the absence of either the 'ideal' test or a scheduled proof test one could resort to the philosophy of acceptance of safeguards system capability on the basis of its response to a genuine random process demand. This philosophy is one which should be approached with much caution, since automatic trip action is very often possible even though dangerous unrevealed failures are present and the protective system therefore remains in a definable failed state. This may be illustrated by reference to, say a temperature, pressure or other analogue input subsystem whose trip level has drifted through component failures or human error to a higher trip datum than that actually demanded for trip action. It follows that a process demand will exceed the permissible set value before a trip is initiated, and hence the protective system is in reality in a failed state even though a trip may have fortuitously taken place. A further instance which would constitute a very hazardous process situation would be with the same failures being present but subjected to a very slow process transient which occupies the band between demanded and actual trip levels over some prolonged period. In such an instance the process could be said to be without protection.

In conclusion, for the above reasons, proof testing is a mandatory requirement and is almost always, because of process restrictions, carried out by means of simulation at the input and output subsystems of the safety system.

4.3 PROOF-TESTING PROCEDURES

A protective system consists of three subsystems:

- input system
- logic system
- shutdown (output) system.

4.3.1 Input system

The input system receives measured data from the plant stream and compares it with a preset trip or alarm limit. It consists essentially of a sensor which is taken to a measuring system which features an adjustable trip datum. When the trip level is reached, the input system generates an output signal which is routed to the logic. It is normally assumed, possibly subject to verification, that the sensor is located at a representative point in the process. The sensor output may be of analogue or digital form, and therefore simulation of an input test signal must be representative of these two possibilities. The following guide-lines state in general terms the requirements of a representative proof test procedure.

(a) Analogue testing

For purposes of analogue testing, a simulated varying input signal not exceeding the set trip level value should be applied over the band of values from normal operation to trip datum at the most pessimistic dynamic rate of rise or decrease which could arise within the process. Additionally scale calibration in terms of accuracy and repeatability at the normal working and trip points should be verified and shown to be within the bounds of acceptability. At the trip level the output signal to the logic should be seen to be initiated.

(b) Digital testing

A digital process input signal may be derived from, say a limit switch denoting a position or state, or a pressure switch denoting a process parameter limit. In the latter instance a simulated digital signal would not be ideally representative of the process condition since its response time of operation would depend on a dynamic analogue pressure change which involves delay from normal operating level to pressure switch output contact operation. Therefore in general terms, where a digital output is derived from a dynamic analogue parameter, the provision of a suitable analogue testing facility at the process sensor would need to be considered.

4.3.2 Logic system

Safety logic may take a number of forms, from hardwired relay or solid state systems to computer software, whose function is to provide a decision-making interface between the safety system input and output subsystem. Logics may be designed for either singular channel or various degrees of majority voting. In practice, logic systems are found to be the most reliable areas of safeguard systems. Where guard line action circuits are used it is necessary to ensure that the ends of the individual guard lines are always physically separated at their termina-tions by an adequate distance in order to avoid accidental short circuits, which would hold a guard line permanently energized. When proof testing safety logic it is possible in many cases to combine the test with that of the input system, i.e. to apply a simulated input signal and to observe subsequent action of the logic output.

In many cases, particularly on continuous processing plants, the logic output may have bypass facilities arranged such that the automatic shutdown system will remain in an energized state in order to keep the plant running during the period of test. It follows therefore that in such instances particular attention should be drawn to the possibility of human error in resetting the logic override after test. Additionally where guard lines are used, facilities should be present which enable the tester to verify operation of interlocks which may be present at the guard line ends. A means is also advisable to enable the required response time interval from input signal to operation of the logic output to be verified.

4.3.3 Shutdown (output) system

Shutdown systems are most often automatic in operation since many process dynamics preclude a human operator response to an alarm state. Where process dynamics are sufficiently slow, the human operator may take the place of the logic system and hence would take on the role of an interface between the input alarm and shutdown systems.

Shutdown systems take many forms such as automatic process valves, heaters, pumps, cooling systems etc. In the area of safeguards, delay times, e.g. automatic valve closures, are usually significant and therefore response times need to be verified when proof tests are carried out. The overall safeguards response from trip level to complete shutdown system operation is a very important part of any proof test since excessive times would indicate that equipment at subsystem or component level somewhere in the safety system chain is moving into the wearout phase C of the bathtub curve (Figure 1.1).

4.4 SYSTEM PROOF TESTING

Proof testing of safety systems is necessarily carried out on both continuous and batch chemical processing systems and is particularly relevant to the partly reversible state discussed previously in Chapter 3. Proof testing is normally carried out on-line in the case of continuous processing and usually off-line in batch processing.

4.4.1 On-line testing

For a protective system, which may be single or multichannel in form, the statistical downtime or FDT comprises three elements:

1. statistical failed time due to random unrevealed equipment faults;
2. time taken to carry out the proof test;

3. failed proof test intervals due to human error whereby the maintenance operator restores the system into service with an unrevealed
dangerous fault generated.

The FDT due to 1 is $\theta\tau/2$ (exponential distribution). The FDT due to 2 is
t/τ, where t is the test duration. The FDT due to 3 relates to human
error in testing and is based on the probability of a failed interval on
the basis that if one failed interval occurred out of n_p proof tests then
the contribution per proof test interval would be $1/n_p$.

The overall FDT for on-line testing is therefore given as

$$\mu_0 = \frac{\theta\tau}{2} + \frac{t}{\tau} + \frac{1}{n_p}$$

These three elements are represented graphically in Figures 4.1(a)–(c).
Combining Figures 4.1(a)–(c) gives Figure 4.1(d) which shows that
there is an optimum value of proof test interval τ_{op} for a minimum
FDT μ_0. This will be given by

$$\frac{d\mu_0}{d\tau}\ \frac{\theta\tau}{2} + \frac{t}{\tau} + \frac{1}{n_p} = 0$$

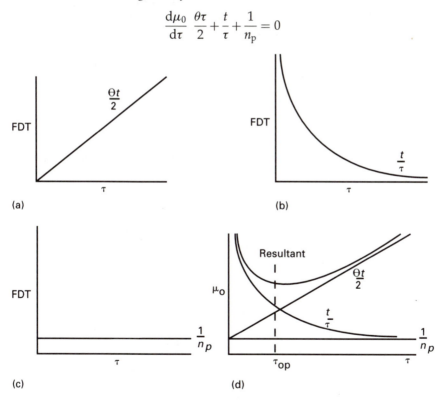

Figure 4.1 (a)–(c) The three FDT elements; (d) resultant FDT from combination
of (a)–(c).

Thus

$$\frac{d\mu_0}{d\tau} = \frac{\theta}{2} - \frac{t}{\tau^2} = 0$$

hence

$$\frac{\theta}{2} = \frac{t}{\tau^2}$$

for minimum μ_0. Hence

$$\tau_{op}^2 = \frac{2t}{\theta}$$

therefore

$$\tau_{op} = \sqrt{\frac{2t}{\theta}}$$

It is generally found in practice that the testing time contribution to system FDT only becomes significant when proof test intervals are less than six weeks. However, Figure 4.1(d) shows that reduction of the test interval beyond τ_{op} will adversely affect the overall FDT, this being due to the increasing prominence of the testing time contributor. Hence as $t \rightarrow \tau$ the operating plant system progressively loses protection from the safety channel under test.

4.4.2 Off-line testing

In this mode, tests are normally carried out during process shutdown, therefore the test duration does not represent a contribution to the overall FDT μ_0. It follows that the elements of μ_0 will now be:

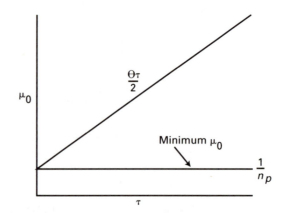

Figure 4.2 Resultant FDT from combination of $\theta\tau/2$ and $1/n_p$.

- $\theta\tau/2$ (due to random dangerous failures)
- $1/n_p$ (due to dangerous human error faults).

These are combined by Figures 4.1(a) and (c) to give Figure 4.2, which shows that as the test interval is reduced the FDT approaches the limit set by human error.

4.5 PROOF TESTING OF THE SINGLE SAFETY CHANNEL

The optimization of proof testing frequency enables the protective system maximum failure probability and hence FDT to be maintained at desired levels which are compatible with the plant protective require-ment. At the time of each test the failure probability function recom-mences from a zero value providing that renewal has taken place when necessary and the system is returned to its initial fault-free state in the useful life phase. This is illustrated in Figure 4.3, which shows the cyclic probability experienced by the simple exponential cumulative failure probability function $P = 1 - e^{-\theta t}$.

Referring to Figure 4.3 the maximum failure probability for a single safety channel is limited on a cyclic basis to a maximum value P_{max} by carrying out proof tests at equal intervals τ corresponding to times t_1, t_2, t_3 etc. The maintenance of P_{max} is conditional on the system being restored to its original fault-free operational state at the completion of each test, i.e. restoration of full capability. This requires that any signifi-cant departures from the confirmed and documented acceptance states at the previous test will be rectified by maintenance adjustments and replacements where necessary of components which are moving into the wearout phase. Replacements when carried out must always be subject to a reasonable soak test period in order to eliminate early life manufacture-based failures.

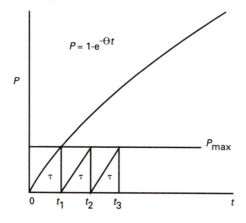

Figure 4.3 Cumulative function showing cyclic failure probability.

5

Data and mean fractional dead times

5.1 DATA FOR FAILURE PROBABILITY EVALUATIONS

A simplified safety system may typically comprise three non-redundant subsystems consisting of input, logic interface and shutdown systems. In safety systems associated with higher risk scenarios it is normally the practice to invest redundancy with possible diversity into one or more of the three subsystems.

Assessment of a safety system must firstly consider the individual subsystem channels which constitute the input to the composite system in terms of their respective failure probabilities. Assembly of these subsystem channel evaluations enables the overall safety system reliability to be determined. The objective is to derive a cumulative failure probability expression, typically $P = 1 - e^{-\theta t}$, assuming failures are exponentially distributed. In order to derive the expression it is necessary to establish a failure probability for the channel under assessment based on a proof test interval which would be compatible with both the safety requirement and on-site maintenance practices. The latter is seen to be flexible and predominantly under the control of site operational management and hence an optimum value should be more easily attainable. The failure probability of a single system, subsystem or component, based on a given proof test interval is dependent on failure rates influenced by manufacturing, design limitations and working environments, and can therefore present some difficulty when seeking quantification. There are three sources of subsystem failure data, namely empirical field data drawn from site maintenance records, failure mode and effect analyses (FMEAs) and reference to established data bank sources.

5.1.1 Empirical field data

Most organizations prefer, usually on grounds of cost and possible lack of expertise, not to establish an in-house data retrieval regime. However, since failure rates are significantly influenced by on-site working factors, empirical data are regarded as the most representative, and therefore use of data from this source is encouraged. Where specific collection regimes are not implemented it is most often possible to derive data from maintenance records whether breakdown or planned modes of working are employed. Maintenance records should indicate dates and reasons for failure and the subsequent measures taken to effect repair. A failure rate for a given subsystem will be significantly dependent on the mode of maintenance which is carried out. It is generally found that failure rates are somewhat worse under breakdown regimes rather than the planned type, this being due to preventive maintenance practices which tend to reduce the incidence of failures.

Data drawn from maintenance sources are usually in a total form, i.e. they comprise both the safe and dangerous modes and therefore judicious judgements need to be exercised in order to derive the dangerous modes of failure. This need is considerably simplified where planned maintenance is used, i.e. revealing and repair of dangerous unrevealed failure modes through a maintenance programme which, with advantage, should include regular proof testing.

5.1.2 Failure mode and effect analyses

More commonly known as FMEAs, such analyses are in reality a technical audit of a system or subsystem at the component level. For example it may be necessary to determine the dangerous failure of, say a temperature trip amplifier. When applying an FMEA to such a system the assessor will consider all possible modes of failure for each component and for each possibility tabulate its effect on the system output, i.e. dangerous or safe. To complete the audit the dangerous failure rates are summed in order to provide the overall system dangerous rate. The system safe failure mode will also, through similar treatment, be available to the assessor.

The FMEA is a method of failure rate determination which is to be recommended when new or novel types of equipment are being used and there is no adequate history of failures over an extended process experience. The main disadvantage of an FMEA approach lies with the problem of applying representative environmental factors to each of the components such as ambient temperature, humidity, loading stress factors etc. A further disadvantage is the higher relative cost of carrying out the analysis in that it is time consuming and demands a specialist knowledge reliability-based approach to the system and therefore is not likely to be within the expertise of the user.

5.1.3 Data banks

The establishment and operation of a data bank service is costly in respect of operation and time involved in retrieval and analysis of collected data. Because of these factors the establishment of an in-house data bank can be largely prohibitive, and therefore the general user's only recourse is to consider possible club membership of an organization such as AEA Technology as a typical example. Obtaining data in this way can also be expensive to the user since it normally involves club membership charges and further charges per item of data. These can sometimes be offset by a reciprocal exchange agreement whereby the data bank authority will collect data from the user's specific organization. In this way the pooled data will refer to a range of environmental factors with the advantage that a user can expect to obtain information from other users which will fit his or her particular use and environment.

5.1.4 Summary of advantages and disadvantages

(a) Empirical field data

- These are the most reliable data since they reflect the user's specific environment and site practices.
- They are normally available and less costly through comprehensive maintenance records.
- Data confidentiality can be maintained against commercial competitors.
- A degree of expertise is required to interpret data from maintenance records which are basically designed for operational management purposes.
- They are more costly if collected through a dedicated retrieval regime.
- Data will most likely refer to total mode failures. Safe and dangerous modes need to be judiciously evaluated.

(b) Failure mode and effect analyses

- They will indicate weak areas of reliability in the subsystem which may promote an improved user specification.
- They normally require outside expertise to carry out an analysis.
- They are costly and time consuming.

- There are difficulties in applying specific site environmental factors to an analysis particularly in relation to breakdown or preventive maintenance.

(c) Data banks

- They cover a wide field of data types drawn from a large cross-section of users.
- Data can usually be related to users' specific needs and conditions.
- Club membership can be costly.
- Communication problems can arise when directing specific data requirements through a third party not conversant with the particular site on which the data will be used.
- Delay times from request to receipt of data may be excessive.
- The user is normally expected to reciprocate data transfers. This could represent difficulties in confidentiality.

5.1.5 Failure rate presentations

Failure rate data required to carry out system assessments are always referred to a single item or subsystem. Engineering safety assessments normally use the format of **faults per annum** for each of the appropriate items or subsystems which constitute the system mathematical model. Data banks may present data in any one of three formats:

- number of failures of a single item per annum;
- mean time between failures of a single item in years (MTBF);
- number of failures of a single item per 10^6 hours.

For a population of 500 samples with a total of ten failures over a period of two years:

Faults per annum

$$= \frac{\text{Total number of failures}}{\text{Total operating time of the population samples}}$$

$$= \frac{10}{500 \times 2} = 0.01 \text{ faults per annum}$$

Mean time between failures

$$= \frac{\text{Total operating time of the population samples}}{\text{Total number of failures}}$$

$$= 100 \text{ years (i.e. reciprocal of faults per annum)}$$

Failures per 10^6 hours

$$= \frac{\text{Failures per annum}}{8760} \times 10^6$$

$$= \frac{0.01}{8760} \times 10^6 \text{ (where } 8760 = \text{number of hours in one year)}$$

$$= 1.142 \text{ faults per } 10^6 \text{ hours}$$

5.2 SYSTEMS MEAN FRACTIONAL DEAD TIMES

Mean fractional dead times (FDTs) describe in engineering terms the reliabilities or unavailabilities of safety systems in both single-channel and composite multichannel voting configurations. The FDT may be more fully defined as:

- the failure probability on demand of a safety system; or
- the average failure probability of a safety system over a given period of time or more specifically the proof test interval τ.

The mean fractional dead time is a mandatory requirement in specifying and maintaining a given safety system to meet a required reliability safety criterion. There are two classes of safety systems:

- non-redundant single-channel system;
- redundant multichannel system.

5.2.1 Non-redundant single-channel system

To evaluate the FDT of a single system safety channel or subsystem based on an exponential distribution of failure it is first necessary to specify an assessed failure rate and also a preferred proof test interval. Evaluation necessitates mathematical treatment in terms of the pertinent overall cumulative probability function which in the single-system case would be of the form $P = 1 - e^{-\theta t}$, where t is time and θ the failure rate. The method of evaluation calculates the mean height P_{av} of the cumulative function over the proof test interval τ. Figure 5.1 illustrates the procedure.

$$P_{av} = \text{Mean height of function } P = \text{FDT}$$

$$\text{FDT} = \frac{\text{Area under curve over test interval } \tau}{\text{Test interval } \tau}$$

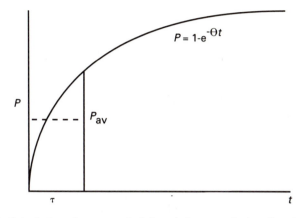

Figure 5.1 Calculating the mean height of the cumulative function over the proof test interval.

$$\text{FDT} = \frac{1}{\tau} \int_0^\tau 1 - e^{-\theta t}\, dt$$

$$= \frac{1}{\tau} \left[t - \frac{e^{-\theta t}}{-\theta} \right]_0^\tau$$

$$. = \frac{1}{\tau} \left[\left(\tau + \frac{e^{-\theta \tau}}{\theta} \right) - \left(0 - \frac{e^0}{-\theta} \right) \right]$$

$$= \frac{1}{\tau} \left(\tau + \frac{e^{-\theta \tau}}{\theta} - \frac{1}{\theta} \right)$$

$$= 1 + \frac{e^{-\theta \tau}}{\theta \tau} - \frac{1}{\theta \tau}$$

i.e.

$$\text{FDT} = 1 + \frac{1}{\theta \tau}(e^{-\theta \tau} - 1) \tag{5.1}$$

(a) Procedure for FDT approximation when $\theta \tau < 0.1$

Using the expansion

$$e^{-x} = 1 - x + \frac{x^2}{2!} - \frac{x^3}{3!} + \dots$$

equation 5.1 expands to

$$1 + \frac{1}{\theta\tau}\left(1 - \theta\tau + \frac{\theta^2\tau^2}{2!} - \frac{\theta^3\tau^3}{3!} + \ldots - 1\right)$$

$$= 1 - 1 + \frac{\theta\tau}{2} - \frac{\theta^2\tau^2}{6} \ldots$$

$$= \frac{\theta\tau}{2} - \frac{\theta^2\tau^2}{6} \ldots$$

Therefore if $\theta\tau < 0.1$,

$$\mathrm{FDT} = \frac{\theta\tau}{2}$$

(b) Summary

For a single system when $\theta\tau \gg 0.1$,

$$\mathrm{FDT} = 1 + \frac{1}{\theta\tau}\left(e^{-\theta\tau} - 1\right)$$

For a single system when $\theta\tau < 0.1$,

$$\mathrm{FDT} = \frac{\theta\tau}{2}$$

5.2.2 Redundant multichannel system

Simple redundancy

Redundant systems comprise two or more protective channels arranged such that shutdown or alarm action is initiated on the evidence of a majority vote from the multichannel system. The simplest vote system consists of two channels known as a 'one-out-of-two' success voting logic which implies that both must fail in order to negate the safety function.

Consider two channels A and B arranged in a one-out-of-two success voting logic. Let the respective channel failure rates be θ_A and θ_B and the common proof test interval be τ. Therefore

$$\text{Failure probability of the two-channel system} = P_A P_B$$

If $\theta\tau < 0.1$ then

$$P_A P_B = \theta_A \theta_B t^2$$

Therefore

$$\text{FDT of } A \text{ and } B = \frac{1}{\tau} \int_0^\tau \theta_A \theta_B t^2 \, dt$$

$$= \frac{1}{\tau} \left[\frac{\theta_A \theta_B t^3}{3} \right]_0^\tau$$

$$= \frac{1}{\tau} \left(\frac{\theta_A \theta_B \tau^3}{3} \right)$$

$$= \frac{\theta_A \theta_B \tau^2}{3}$$

If $\theta_A = \theta_B$, then

$$\text{System FDT } \mu_D = \frac{\theta^2 \tau^2}{3}$$

(b) Majority voting

Simple redundant systems comprise two or more identical protective channels arranged such that any one channel out of the total of n such channels will provide the desired protective function. Simple redundancy carries with it a significant advantage in safety system reliability but unfortunately is accompanied by a severe penalty in the form of an increased incidence of spurious process trips. Such trips are caused by revealed safe failures in each of the protective channels which often result in production losses as a result of unscheduled plant shutdowns. For a system of n channels in simple redundancy the overall system reliability is improved by a power of n such that if P is the failure probability of a single channel then the failure probability of the overall n-channel system will be P^n. Since P is fractional, P^n represents a significant improvement in reliability of the protective system.

However, since each protective channel will have a revealed failure rate of, say θ, the spurious plant tripping rate will be degraded by a factor of n to give a protective system spurious rate of $n\theta$.

In order to reduce the incidence of spurious trips to an acceptable level whilst at the same time meeting a given safety reliability criterion, majority voting is used. To achieve this the simple redundant success logic of 'one out of n' is modified to a higher form, namely 'm out of n'.

Success and equivalent failure logics may be conveniently transposed through the expression $r = n - m + 1$ where:

- r is the minimum number of channels required to fail in order to inhibit the overall system;
- n is the total number of channels which constitute the composite safety system;
- m is the minimum number of working channels required to give overall system success.

Table 5.1 lists the more commonly used voting success logics and their equivalent system failure logics. Table 5.2 lists mean fractional dead times for a range of r/n system failure logics which are subject to simultaneous proof testing. These are derived from truth tables and Boolean-related expressions which are discussed in Chapter 6.

In practice it is beneficial to carry out proof testing on a staggered basis. Such testing will realize two important benefits.

Table 5.1 System success/failure voting logic equivalents

Success voting logic	Equivalent failure logic
1 out of 3	3 out of 3
2 out of 3	2 out of 3
2 out of 4	3 out of 4
3 out of 4	2 out of 4

Table 5.2 Simultaneous proof testing

			n		
r	1	2	3	4	5
1	$\dfrac{\theta t}{2}$	θt	$\dfrac{3\theta t}{2}$	$2\theta t$	$\dfrac{5\theta t}{2}$
2	–	$\dfrac{\theta^2 t^2}{3}$	$\theta^2 t^2$	$2\theta^2 t^2$	$\dfrac{10\theta^2 t^2}{3}$
3	–	–	$\dfrac{\theta^3 t^3}{4}$	$\theta^3 t^3$	$\dfrac{5\theta^3 t^3}{2}$
4	–	–	–	$\dfrac{\theta^4 t^4}{5}$	$\theta^4 t^4$
5	–	–	–	–	$\dfrac{\theta^5 t^5}{6}$

- The composite safety system FDTs are improved. Table 5.3 lists the improvement factors comparing simultaneous proof testing with symmetrical staggered testing.
- Staggered proof testing is an important factor in reducing the incidence of common-mode failures.

Table 5.3 Improvement factors – symmetrical proof testing

	n				
r	*1*	*2*	*3*	*4*	*5*
1	1	1	1	1	1
2	–	1.6	1.5	1.45	1.43
3	–	–	3.0	2.67	2.50
4	–	–	–	6.12	5.21
5	–	–	–	–	13.17

6

Logical network principles

6.1 INTRODUCTION

Logical networks are recognized and widely used as powerful tools in deriving mathematical models of systems in terms of risk and reliability. Such networks are commonly known as **fault trees** and are built up from a variety of symbolic logic gate elements. Each gate element features multiple input states which, if present concurrently, will result in an output from the gate. The fault tree itself progresses from primary events to a resultant and final end event.

Logic gates are type dependent in their specific uses and are subject to certain rules in respect of input progressions through the gates themselves, as well as integration of their outputs with those of other gates. Gates fulfil two specific roles in that they provide the means of combining:

- binary variables in fault trees;
- probabilistic quantities in mathematical models.

Safety engineering calls for the expression of safety system probability in terms of failure rather than success in order to promote compatibility with:

- the assessment of risk which is based on the product of dangerous demands and failure probability of the protective system;
- the failures of interest in safeguard systems, being those which negate safety actions.

6.2 BINARY VARIABLES

Logical networks, i.e. fault trees, are concerned solely with processing binary variables. Such variables can have only two states which represent false or true hypotheses, and when combined yield a finite number of possible output combinations which are called sets. Combinations take place through logic gates and the mathematics which is applied is a two-state binary algebra known as **Boolean algebra**. The purpose of this book is to present a working appreciation of safety engineering principles rather than a mathematical treatise, hence the Boolean relationships to be introduced are intended to be sufficient for the stated purpose of fault-tree analysis.

6.2.1 Boolean algebra

Boolean algebra considers only two states, numerically represented by 0 and 1, which by convention represent false or true hypotheses respectively.

The algebra, when applied to practical safety engineering is most conveniently confined to two mathematical operations, namely **AND** and **OR** such that:

- AND implies multiplication and is symbolized by '&' (though the sign can be omitted in equations);
- OR is symbolized by '+', but note that the sign does not imply algebraic addition.

In order to clarify the binary hypotheses of 0 and 1 it is convenient to consider the following propositions whereby a hypothetical state A is either true or false.

- The A state being true or present is equivalent to 1 and is simply stated as A.
- The A state being false or not present is equivalent to 0 and is simply stated as $\bar{A}$ (bar A).

Applying the above principles,

$$A + \bar{A} \quad (\text{i.e. } A \text{ or } \bar{A}) = A$$

(A is present since 1 OR 0 = 1, the priority state being 1);

$$A\bar{A} \quad (\text{i.e. } A \text{ AND } \bar{A}) = 0$$

(since $0 \times 1 = 0$);

$$AA \quad (\text{i.e. } A \text{ AND } A) = A$$

(since $1 \times 1 = 1$).

6.2.2 Boolean relationships

The following relationships represent the essential equations necessary for evaluating fault trees.

$$A + A = A \tag{6.1}$$

$$A + \bar{A} = 1 \tag{6.2}$$

$$AA = A \tag{6.3}$$

$$A\bar{A} = 0 \tag{6.4}$$

$$A + AB = A(1 + B) = A \tag{6.5}$$

$$A + AB + ABC + ABCD = A(1 + B + BC + BCD) = A \tag{6.6}$$

$$A + \bar{A}B = A + B \tag{6.7}$$

The equation $A + \bar{A}B$ can be verified by assigning values of 0 and 1 successively to A. Alternatively the identity can be conveniently proved by means of a truth table approach which is discussed later in this chapter.

6.3 LOGIC GATES – SYMBOLIC NOTATIONS

Fault trees may be constructed from three gate types, namely 'AND' gates, 'OR' gates and 'voting' gates, which are shown in Figures 6.1(a)–(c) and are drawn using the generally accepted symbolic notations. These gate symbols have a serious disadvantage in that engineering professions not fully conversant with reliability or safety engineering have difficulty in recognizing gate function types based on shape only. It is therefore strongly recommended that all gates should be shown as $\frac{3}{8}$ inch (10 mm) diameter circles with the function type clearly written within so as to convey instant recognition. The recommended alternatives are therefore offered in Figures 6.2(a)–(c) and will be used henceforth.

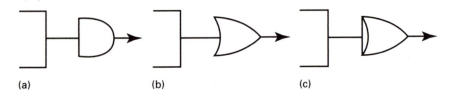

(a)　　　　　　　　(b)　　　　　　　　(c)

Figure 6.1 Symbolic representations of: (a) AND gate; (b) OR gate; (c) voting gate.

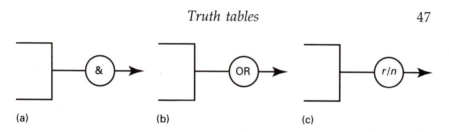

Figure 6.2 Alternative written representations: (a) AND gate; (b) OR gate; (c) voting gate.

6.4 TRUTH TABLES

Truth tables are simply a means of tabulating all the possible combinational states of element groups in order to realize system expressions which can be either deterministic or probabilistic. Boolean algebra is deterministic in that it postulates binary states such that a condition is either present or not present, and hence relates to discrete probabilities of 1 and 0. In the practical engineering world it is not possible to have knowledge of a definite elemental state but rather its probability of being 'present' or 'not present', which implies a probabilistic hypothesis of being somewhere between 1 and 0.

To construct a truth table it is necessary to list in appropriate columns the defined elemental states such that taken together they express all the possible resultant system states which they produce. Truth tables may be drawn up for any number of elements but become somewhat large above six. Table 6.1 shows a truth table for three elements A, B and C. When considering Boolean relationships the respective element states are shown as either:

$$\text{Present} = A \text{ or } \quad \text{Not present} = \bar{A}$$

Table 6.1 Truth table – three elements in present/not present logic

A	B	C
A	B	C
A	B	$\bar{C}$
A	$\bar{B}$	C
A	$\bar{B}$	$\bar{C}$
$\bar{A}$	B	C
$\bar{A}$	B	$\bar{C}$
$\bar{A}$	$\bar{B}$	C
$\bar{A}$	$\bar{B}$	$\bar{C}$

It is also important to define whether the 'present' state relates to failure or success. In Table 6.1 'present' refers to success, hence 'not present' refers to the presence of failure. Evidence of correct compilation is verified whereby the table commences with the top row showing all three element states as 'present' whilst the bottom row shows all the three states as 'not present'. The nomenclature of Table 6.1 is most convenient for the Boolean deterministic approach which is pertinent to fault trees.

Mathematical modelling, which is dealt with in Chapter 8, also relies on the same truth table in order to derive probabilistic expressions relating to a particular model. For convenience, Table 6.2 suggests a recommended modification of the Boolean-based table (Table 6.1) in order to promote conformity with the probabilistic stage requirements. Assuming a failure-based philosophy, the table shows each element state as either working (W) or failed (F).

Table 6.2 Truth table – three elements in working/fail logic

A	B	C
W	W	W
W	W	F
W	F	W
W	F	F
F	W	W
F	W	F
F	F	W
F	F	F

6.5 COMBINATIONS THROUGH LOGIC GATES

Truth tables enable outputs from logic gates to be derived from the input element states. The following considers simple two-element system inputs in respect of AND and OR gates. Similar treatment is also illustrated for a simple two-out-of-three element voting logic gate. It should be noted that in the following truth tables '1' denotes 'present' whilst '0' denotes 'not present'.

6.5.1 The AND gate

Table 6.3 represents the truth table for a two-element input AND gate.

Table 6.3 Truth table – two-element input AND gate

Row no.	A	B	AB
1	A	B	1
2	A	$\bar{B}$	0
3	$\bar{A}$	B	0
4	$\bar{A}$	$\bar{B}$	0

Row 1 of the table shows that an output is only present when both *A* AND *B* are present. The two-element Boolean logic is given in Figure 6.3.

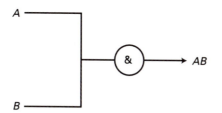

Figure 6.3 Boolean representation of a two-element AND gate.

6.5.2 The OR gate

Consider the two quantities *A* and *B* arranged in an OR logic configuration (Table 6.4). Rows 1–3 show that an output is present for each of the

Table 6.4 Truth table – two-element input OR gate

Row no.	A	B	A + B
1	A	B	1
2	A	$\bar{B}$	1
3	$\bar{A}$	B	1
4	$\bar{A}$	$\bar{B}$	0

pertinent combinational states of *A* and *B*. Hence evaluation in Boolean terms concludes that *A* OR *B* is present when conditions called for by rows 1–3 are satisfied, i.e.

(A AND B)   OR   (A AND $\bar{B}$)   OR   ($\bar{A}$ AND B)

$$= AB + A\bar{B} + \bar{A}B$$

$$= A(B + \bar{B}) + \bar{A}B \quad \text{(note } B + \bar{B} = 1)$$

$$= A + \bar{A}B$$

$$= A + B$$

The logic for the two-element OR gate is given in Figure 6.4.

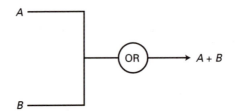

Figure 6.4 Boolean representation of a two-element OR gate.

6.5.3 The majority voting gate

Output expressions for majority voting gates are derived from truth tables using an identical approach to that previously described for AND and OR gates. One of the most common and simple forms met in safety engineering is that of the 'two-out-of-three' (2/3) logic. This implies, by reference to Table 5.1, that for success, any two out of three safety channels must operate in order to produce a protective action from the safety system. Conversely, this also implies that any two out of the three safety channels must fail in order to negate output protective action. The truth table for three elements A, B and C will now be constructed, but note that a similar procedure would also be applicable for other voting logics (as illustrated in Table 5.1).

Table 6.5 represents the relevant 2/3 truth table, noting that 'W' depicts a working or healthy state and 'F' denotes a failed state. It should also be emphasized that since failure is the consideration, the presence of failure will be pertinent to '1' or the 'present' state whilst success will be pertinent to the '0' or 'absent' state. Hence

Channel A failed $= A$

Channel A not failed $= \bar{A}$

In Table 6.5, rows 1–4 list the logic failure states which, if present, will result in overall protection system failure. System failure is therefore given by row 1 OR row 2 OR row 3 OR row 4.

Table 6.5 Truth table – 2/3 voting logic gate

A	B	C	Logic output	Row no.
W	W	W	$\bar{A}\bar{B}\bar{C}$	
W	W	F	$\bar{A}\bar{B}C$	
W	F	W	$\bar{A}B\bar{C}$	
W	F	F	$\bar{A}BC$	1
F	W	W	$A\bar{B}\bar{C}$	
F	W	F	$A\bar{B}C$	2
F	F	W	$AB\bar{C}$	3
F	F	F	ABC	4

Hence the Boolean equation for 2/3 failure is

$$\bar{A}BC + A\bar{B}C + AB\bar{C} + ABC$$

$$= BC(\bar{A} + A) + A\bar{B}C + AB\bar{C}$$

$$= BC + A\bar{B}C + AB\bar{C}$$

$$= C(B + A\bar{B}) + AB\bar{C}$$

$$= C(B + A) + AB\bar{C}$$

$$= CB + CA + AB\bar{C}$$

$$= CB + A(C + B\bar{C})$$

$$= CB + A(C + B)$$

$$= CB + AC + AB$$

$$= AB + AC + BC$$

The logic for 2/3 failure is given in Figure 6.5.

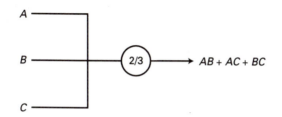

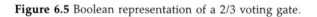

Figure 6.5 Boolean representation of a 2/3 voting gate.

6.6 DIMENSIONAL CRITERIA

Observance of dimensional criteria at logic gates is a strict requirement when formulating logical networks in terms of fault trees and probabilistic mathematical models. Networks comprise only two quantities:

- rates;
- probabilities.

Rates refer to events per unit time, e.g. faults per annum, which in dimensional terms represents a pure number divided by time. Hence the dimension of rate is that of inverse time, denoted by $[T^{-1}]$ where T is time.

Probabilities refer to the chance of an event occurring and therefore in dimensional terms is a pure number signified by [1]. The combining of rates and probabilities at logic gates must satisfy the mandatory requirement of a single dimensional output from any given gate. Combinations at Boolean OR and AND gates are now considered in terms of dimensional analysis.

6.6.1 OR gate

Figures 6.6(a)–(c) refer to OR gates with all possible combinations of rates and probabilities. The analysis concludes that rates and probabilities cannot be combined at an OR gate. Figure 6.6 shows that the combination of these two quantities is invalid since it does not produce a singular dimensional output from the OR gate. The combination is also invalid when applied to probabilistic mathematical modelling. This is referred to in Chapter 8.

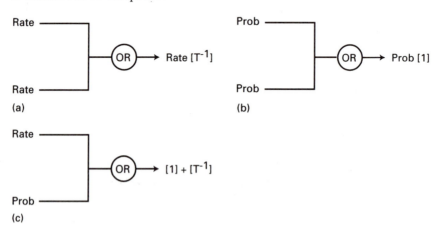

Figure 6.6 OR gates showing all possible combinations of rates and probabilities.

6.6.2 AND gate

Figures 6.7(a)–(c) consider all the possible input combinations of rates and probabilities. It is concluded that there are no restrictions due to dimensional requirements at Boolean AND logic gates.

It is, however, important to emphasize that Figure 6.7(a), although fully compatible with Boolean logic, is not directly so for probabilistic modelling. This will be discussed in Chapter 8 which deals with mathematical modelling.

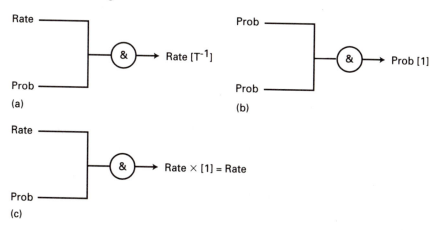

Figure 6.7 AND gates showing all possible combinations of rates and probabilities.

6.6.3 Conclusions – Boolean logic gates

- Input sets which contain both rate and probability elements cannot be combined at Boolean OR gates.
- Rates and probabilities may be combined at AND gates.

7

Fault trees

7.1 INTRODUCTION

Fault trees may be defined as logical networks interconnected by Boolean logic gates which show how multiple basic failure events lead to a single undesirable end event. Reliance can be made on three principal logic gate types, namely 'AND', 'OR' and 'majority voting'.

A given fault tree in its entirety is essentially a composite plant system logical expression in strict engineering terms. It therefore represents a powerful tool in that it provides an essential bridge of understanding between engineers, designers, process plant personnel and the safety and reliability analyst. In the classical situation the analyst would have little or no initial knowledge of the particular plant system and the process system personnel will likewise have little knowledge of the safety and reliability techniques which need to be applied in order to satisfy engineering safety requirements.

Fault trees also yield an important bonus in that they provide a greater degree of understanding of the plant system by all concerned. They also generate an awareness of the significant basic failure events which is advantageous when seeking to achieve a good safety practice philosophy.

Also because of their engineering environment, it follows that process engineers, designers and operational managements are well qualified to carry out fault-tree constructions. The empirical nature of fault trees promotes, with experience, easier compilations, i.e. one becomes more skilful with experience.

7.2 FAULT-TREE SOFTWARE PROGRAMS

There are a substantial number of fault-tree software programs which have been developed and, when used by experienced safety analysts, offer some advantage. However, practically all available programs also incorporate mathematical modelling as part of the software package and hence the description of 'fault-tree program' is in reality a misnomer.

There are two distinct and separate stages in carrying out an assessment. The first is that of deriving the fault tree from systems engineering knowledge with the sole objective of deriving a minimal cut-set listing (sections 7.5 and 7.6). The second stage is that of producing a mathematical model of the system under study by applying the fault-tree cut-set listing as a basis for the model. There is therefore a transition from the deterministic fault tree to that of the probabilistic, mathematical model. The two models differ in that the former is expressed in Boolean algebra, whilst the latter is expressed in terms of traditional algebra.

The compilation and evaluation of fault trees is reasonably straight-forward providing dimensional criteria are observed at all the gates, and system engineering is correctly interpreted. On the other hand, a mathematical model based on a given fault-tree cut-set listing will be subject to a variety of interpretations according to specific requirements of the assessment and it is therefore encumbent on the analyst to correctly identify the pertinent options and to construct the model accordingly. Fault-tree/mathematical model software programs in their use rely entirely on the knowledge and experience of the professional analyst and are therefore to be used with much caution by less-experienced individuals.

The use of fault-tree software is not to be wholly deprecated, but would-be users are strongly recommended to work manually through fault-tree case studies of varying complexities and their subsequent mathematical models. This approach is seen to be very necessary in order to be adequately conversant with the rules governing the techniques of modelling and hence to avoid the many pitfalls which less-experienced engineers and designers may encounter.

As a final comment it may be stated that development of fault-tree software programs has taken place in response to the need for cost savings in terms of assessment time and also to enable engineers etc. to access the benefits of the technology with the minimum of training.

7.3 LOGIC PHILOSOPHY – SUCCESS OR FAILURE

Fault trees may be drawn up in terms of either success or failure. Although both types of logic are acceptable there is merit in employing failure logic on the basis that process safety and reliability studies are directly related to failures, i.e. things which 'go wrong'. Failure mode philosophy is therefore recommended and henceforth will be the basis

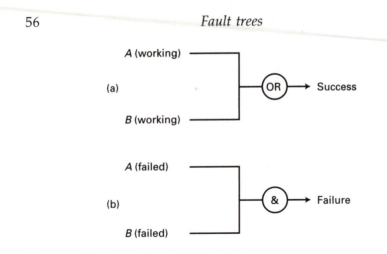

Figure 7.1 (a) Success logic; (b) failure logic.

of all discussion in the chapters which follow. The relationship between success and failure logic is illustrated by Figures 7.1(a) and (b).

7.4 FAULT TREES AND EVENT TREES

Event trees have already been referred to in section 1.1.4. Fault trees and event trees are complementary logic networks, but in order to clarify their differences their respective functions will now be discussed.

7.4.1 Fault trees

A fault tree commences with primary failure events which are combined through logic gates in order to culminate in a single undesirable end event. The nature of the end event, depending on the specific assessment requirement, will be any one of a whole range of outcomes which typically could be a runaway process reaction, release of toxic material or an approach to a dangerous state within the process stream. Where safety systems are under assessment, the end event may be loss of protection due to component failures in the equipment or human error in testing or commissioning. Any single end event as given by a fault tree may go on to a multiplicity of possible consequences ranging from the insignificant to major accident scenario. Consequences are the substance of event trees which will now be discussed.

7.4.2 Event trees

An event tree is a logic network characterized by an absence of logic gates and therefore subject only to probabilistic evaluation. The event tree commences with the hazard end event from the associated fault tree and progresses through nodal points which each pose a 'yes/no' question with associated probabilities applied to each of the hypotheses. The end product of the event tree is a range of consequences which may arise from the presence of the single input hazard event. The possible outcomes therefore cover a spectrum from 'no damage' to those of 'major accident'. In practice the event tree is the tool of the hazards specialist, whilst the fault tree is that of the safety and reliability engineer. Figure 7.2 illustrates in schematic form the relationship between the two types of logic and Figure 7.3 typifies a hypothetical fault-tree–event-tree composite logic.

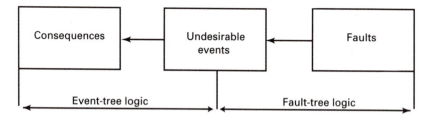

Figure 7.2 Fault- and event-tree composite logic schematic.

7.5 FAULT-TREE RATIONALIZATION

Fault trees are rationalized by means of Boolean algebraic identity expressions which have been previously discussed in Chapter 6. It should be emphasized, since the networks are based at the engineering level, that a given basic failure event can and does often appear more than once in the composite fault tree. At first sight this may seem confusing since such a given failure can only occur once. However, it is pointed out that the fault-tree network is a statement of multiple basic events in the form of element group failures and therefore is in reality a truth-table expression. Repetition of a given basic event is to some extent an endorsement of valid analysis procedure.

The purpose of rationalization, commonly known as a 'Boolean reduction', is to evaluate groups of basic failures known as **cut sets**. These discrete sets each comprise a Boolean AND deterministic expression of a specific group of basic events which, if present concurrently, will bring about failure of the overall system.

7.6 CUT SETS

A cut-set listing is a necessary precursor to the derivation of the ultimate probabilistic mathematical model. As initially derived they may

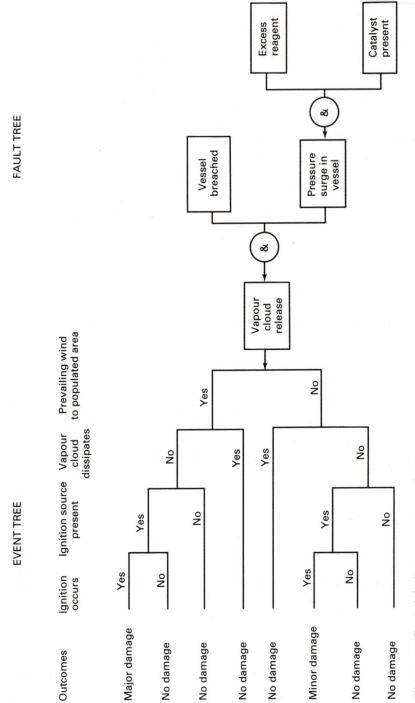

Figure 7.3 Typical fault-tree–event-tree logic.

Table 7.1 Minimal/non-minimal cut-set listing

Set	Cut set	Minimal
1	AB	Yes
2	ACD	No
3	AD	Yes
4	CDE	No
5	ABDE	No
6	DE	Yes

be non-minimal or minimal. Their classifications are such that any cut set which contains a minimal cut set is not minimal. Table 7.1 illustrates this difference. It can be seen that cut sets 1, 3 and 6 are minimal since they do not contain any other cut sets. Cut set 2 is non-minimal since it contains minimal cut set 3. Cut set 5 is also non-minimal since it contains any one of a number of minimal cut sets, namely 1, 3 and 6.

Any one of the six cut sets shown in Table 7.1 will, if present, cause system failure; hence they may be conveniently expressed in terms of Boolean OR logic as shown in Figure 7.4. From Figure 7.4 the Boolean equation is written as follows:

$$AB + ACD + AD + CDE + ABDE + DE$$
$$= AB(1 + DE) + DE(1 + C) + AD(1 + C)$$
$$= AB + DE + AD$$

Hence Figure 7.4 is shown to be equivalent in all respects to that of Figure 7.5. It should be noted that system failure is expressed entirely from the minimal cut-set listing with the non-minimal cut sets being entirely superfluous.

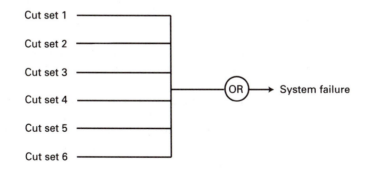

Figure 7.4 Boolean representation of cut sets of Table 7.1.

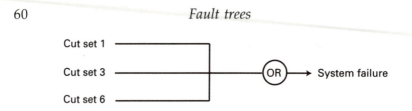

Figure 7.5 Equivalent representation by minimal cut sets of Figure 7.4.

7.7 CATEGORIES OF CUT SETS

Fault trees are constructed in terms of an end event which is either a probability or a rate. Derived cut sets, both minimal and non-minimal, therefore fall into two categories, namely, probability sets and rate or frequency sets.

7.7.1 Probability sets

When the fault-tree end event is a probability, each and all cut sets constitute Boolean AND expressions comprising probability elements only. As an example, for a third-order cut set ABC, each of the elements will represent a probability capable of being expressed as a mean fractional dead time or maximum failure probability according to assessment requirements.

7.7.2 Rate sets

When the fault-tree end event is a rate, i.e. annual frequency of an undesired event, then all cut sets which are each Boolean AND expressions will represent a rate. The most common format of such a cut set would consist of a single rate element and one or more probability elements. Referring to, say, a third-order cut set ABC with the rate element defined as A, the probability elements would be B and C. This is in accordance with dimensional criteria whereby a rate A is multiplied by pure numbers B and C.

It is also possible that a cut set could contain more than one rate element and one or more probability elements which again satisfies dimensional requirements when expressed in Boolean deterministic terms. A final possibility would be where all elements are those of rate with probability elements not present.

7.8 GENERAL PRINCIPLES OF FAULT-TREE CONSTRUCTION

7.8.1 Information requirement

Fault-tree construction is totally dependent on availability of relevant information. In meeting an adequate information requirement there are two principal guidelines which need to be followed. The first is that information in the form of engineering design and operation is made available. Design data normally embrace manufactured components and assembly, including working environments within the equipment itself and the projected operating conditions. It is also necessary to be aware of the designer's operating philosophy and how this complies with the actual operation at the point of application. There is thus a general requirement for information exchange between the reliability engineer, manufacturers, designers and users in order to establish an understanding of the system in terms of the intended function which is pertinent to the assessment.

The second and equally important guideline is that of establishing a realistic level of information sufficient to meet the aims of the assessment. This implies that the study boundaries need to be fully defined and agreed by all interested groups. The attainment of an optimal information level ensures that the assessment undertaking, which can be costly in terms of professional time, is reduced to a minimum. To emphasize this point, awareness of the assessment boundaries avoids time being wasted on areas which are not directly related to the given reliability project. A final report which contains superfluous assessment will only serve to cause confusion and difficulties to the recipients when assimilating the study along with its recommendations and conclusions.

7.8.2 General construction

(a) Engineering expertise

Fault trees, in view of their characteristic logistical interpretation of any given engineered system, should not be regarded as the exclusive domain of the safety and reliability engineer or analyst. System designers and operational managements most often provide a significant system engineering contribution to fault-tree construction. Therefore by extending this function to a working appreciation of dimensional criteria at logic gates followed by subsequent Boolean reduction, a composite fault-tree capability should be reasonably attainable by engineering disciplines generally. There may be some initial difficulties in progression from the single end event to the primary initiating failures. However, with an awareness of the logical approach it is found that, with experience, fault trees become progressively easier to construct.

(b) Fault-tree categories

It is advisable when constructing a fault tree to commence with the undesired end event which has been agreed in discussion with all the contributing engineering groups. Typical examples of an undesired end event could be:

- annual hazard rate;
- failure probability of system.

Fault-tree end events fall into two distinct categories, namely those of hazard frequency and system failure probability. These are illustrated respectively in principle in Figure 7.6.

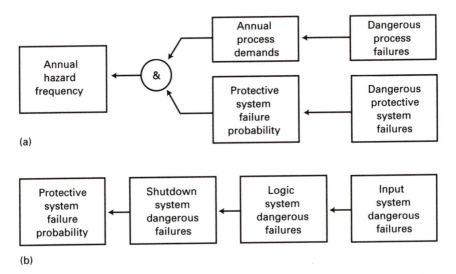

Figure 7.6 Fault-tree schematics: (a) frequency; (b) failure probability.

(c) Formatting principles

Throughout the fault-tree network it is recommended in the interests of clarity and accuracy to insert condition statements at significant points in the logic. The inclusion of such statements, as well as assisting with logical progression, promotes a clearer understanding of how the

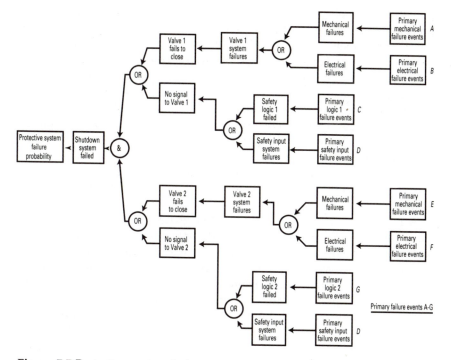

Figure 7.7 Protective system fault tree.

primary failure events give rise to the single end-event state. This tech-nique may be illustrated by reference to a hypothetical safety system which features a shutdown system consisting of two automatic process shutdown valves. The valves are arranged via separate safety system logics to close when protective action is called for by a process demand being present at the input of the safety system. Figure 7.7 shows the general construction procedure which would be used.

(d) Sub-fault-tree networks

Extensive fault-tree networks may suggest difficulties in composite pre-sentation because of ultimate size. Additionally Boolean reductions to minimal cut sets may also become cumbersome because of larger initial expressions. Basically, for these reasons, users tend to view software fault-tree programs as a solution to the difficulties of dealing with larger networks.

Handling larger fault-tree networks can be carried out quite conveni-ently if the network is sectioned into identifiable subsections. When this procedure is applied, which is sometimes necessary, the individual sections are each reduced to the Boolean subequation expressed in minimal terms from which a reduced sub-fault tree can be derived. Assembly of sub-fault trees, along with their respective equations, enables the overall reduced composite fault tree to be conveniently produced with the minimum complications in handling and ultimate resolution.

This technique can be conveniently illustrated by reference to Figure 7.7 which refers to a protective system fault tree expressed in simplified overall terms. In reality the primary failure events A–G would each represent the end event of a sub-fault tree which would later be assembled into the overall composite fault tree shown. The failure event A representing primary mechanical failures attributed to shut-down valve 1 will in fact be constituted by a number of mechanical failures in the process valve itself such as gland stiction, spring failure, valve seat failures etc. The associated three-way electrical pilot valve will also contribute mechanical failures such as stiction, vent port blockage, leakage across valve seats etc. Hence the event A will have an associated Boolean reduced expression which may conveniently be assembled with those of the other end events B–G.

When working with sub-fault trees it is necessary to use identifying input and output symbols, the former being shown on the base fault-tree network and the latter at the output of the sub-fault tree. Referring to the failure event A as an example of their use, the symbols are illustrated in Figures 7.8 and 7.9.

Figure 7.8 Data entry to base fault tree.

Figure 7.9 Sub-fault-tree data termination.

7.9 BOOLEAN REDUCTION

Reduction is a process of writing down for a given fault tree the primary failure events in the form of an overall equation which is characterized by interposing logic gates between the stated primary events and the undesired end event of the composite fault tree. The overall equation when evaluated in accordance with the relationships given previously in section 6.1 will yield the required cut-set listing in minimal form. To introduce the reduction procedure it is convenient to consider the fault tree shown in Figure 7.7 and proceed as follows. Let the protection system failure probability be P_0, then

$$P_0 = [(A + B) + (C + D)][(E + F) + (G + D)]$$

$$= (A + B + C + D)(E + F + G + D)$$

$$= A(E + F + G + D) + B(E + F + G + D) + C(E + F + G + D)$$
$$+ D(E + F + G + 1)$$

$$= (A + B + C)(E + F + G + D) + D \tag{7.1}$$

(note that $D(E + F + G + 1) = D$). P_0 in terms of minimal cut-set listing is given by the expansion of equation 7.1, hence

$$P_0 = AE + AF + AG + AD + BE + BF + BG + BD + CE + CF$$
$$+ CG + CD + D \quad \text{(i.e. 13 sets)}$$

$$= AE + AF + AG + BE + BF + BG + CE + CF + CG + D$$
$$\text{(i.e. 10 minimal cut sets)}$$

In order to establish confidence in any minimal cut-set evaluation it is recommended that several sets consisting of lower and higher orders be selected at random and applied successively to the composite fault tree. This convenient check should verify that they each yield the undesired end-event state by passage through the logical network.

7.10 PROCESS SYSTEM – DEMONSTRATION FAULT-TREE STUDY

7.10.1 Study approach philosophy

The simplified process system shown in Figure 7.10 is introduced as the subject of the following demonstration study for which an engineering level fault tree will be constructed in terms of annual rate of a hazardous end event.

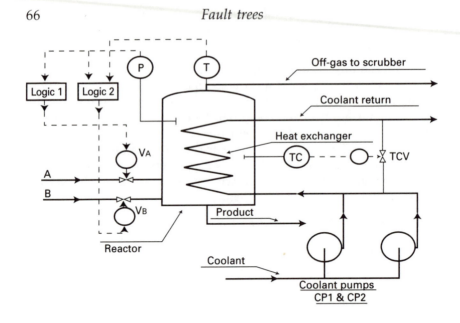

Figure 7.10 Process system – demonstration fault-tree study.

For illustrative purposes the method of construction will rely on the use of sub-fault-tree networks which when assembled into the base fault-tree model will enable the overall or composite fault tree to be expressed in its entirety. It is emphasized that for this particular fault-tree study, division into subsection networks would not be necessary in the practical case. However, for larger networks the employment of this method is recommended and would promote a significant degree of simplication in the eventual Boolean reduction of the overall fault-tree model.

7.10.2 Process description

In the process shown in Figure 7.10 chemicals A and B are passed to a reactor vessel at controlled flow rates. A non-hazardous liquid bottom product is taken off, whilst off-gas is taken from the reactor top to a scrubber system.

The reaction is exothermic and is controlled by coolant flow through the reactor vessel heat exchanger. Coolant is circulated by two coolant pumps CP1 and CP2 which are rated such that any single pump can supply the maximum cooling demand. A breakdown maintenance regime is applied to the pump system.

Reaction vessel temperature is controlled by temperature controller TC which modulates a process valve TCV located in the vessel heat exchanger bypass.

In the event of coolant failure the exothermic heat generated will give rise to a hazardous condition in the reactor which is manifested by a rising pressure in the reactor and increasing temperature of the off-gas.

7.10.3 Process hazard

The hazard is defined as an uncontrolled reaction rate which if allowed to develop fully would result in severe plant damage and possible injury to operating personnel.

7.10.4 Process demands

Process demands are given by those dangerous coolant system failures which may result in a runaway exothermic reaction. Coolant flow can fail due to:

- failures of temperature control system TC whereby the control valve TCV moves to an unscheduled open state;
- loss of both coolant pumps CP1 and CP2.

7.10.5 Protective system

The system consists of an input subsystem, logic interface, and shut-down system.

(a) Input subsystem

The input system features both redundancy and diversity in the forms of vessel pressure and off-gas temperature monitoring.

(b) Logic subsystem

The logic system features redundancy without diversity in the form of two separate logic systems, logic 1 and logic 2. Logic 1 receives the output from reactor pressure sensor P only, whilst logic 2 receives the outputs from both the pressure sensor P and off-gas temperature sensor T. The output from either logic system will fully actuate the shutdown system.

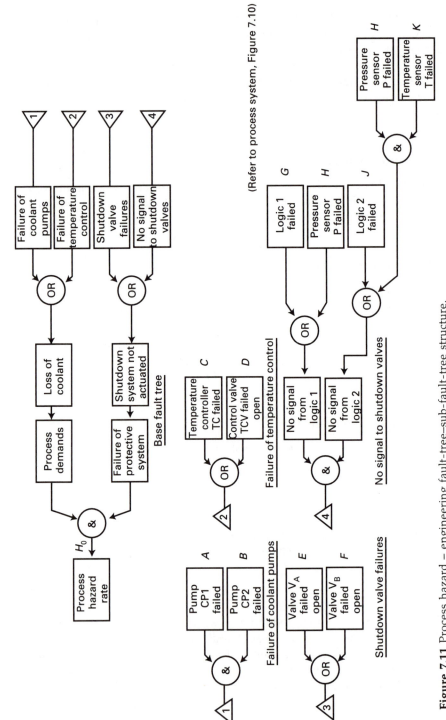

Figure 7.11 Process hazard – engineering fault-tree–sub-fault-tree structure.

(c) Shutdown subsystem

The shutdown system consists of two automatic process shutdown valves V_A and V_B which are arranged to close off flows of reactants A and B respectively in the event of a dangerous process demand. Successful shutdown action is deemed to have taken place only if both valves achieve full closure.

7.10.6 Boolean reduction of composite fault-tree model

Referring to the fault tree shown in Figure 7.11, let

Process system hazard rate $= H_0$

Primary process demands $= A–D$ inclusive

Primary protective system failure events $= E–K$ inclusive

Hence

$$H_0 = [AB + (C + D)][E + F + (G + H)(J + HK)]$$

$$= (AB + C + D)(E + F + GJ + GHK + HJ + HK)$$

Note GHK is not minimal since it contains minimal set HK. Thus

$$H_0 = (AB + C + D)(E + F + GJ + HJ + HK) \qquad (7.2)$$

which is a system factorized Boolean reduced expression. H_0 in terms of minimal cut-set listing is given by the expansion of equation 7.2, hence

$$H_0 = ABE + ABF + ABGJ + ABHJ + ABHK + CE + CF$$

$$+ CGJ + CHJ + CHK + DE + DF + DGJ + DHJ + DHK$$

7.10.7 Conclusions and recommendations

Fault-tree analysis provides a valuable management tool in that it has the capability to reveal either qualitatively or quantitatively the weak areas of the process system being studied. Quantitative analysis will be dealt with in Chapter 8; hence this demonstration study will be confined to conclusions in terms of the qualitative aspect.

There is major advantage in a qualitative assessment approach during early stages of process design, plant modifications or changes in operating philosophy. Qualitative assessment provides valuable guidelines which can significantly reduce the need for costly modifications at later stages of project completion.

(a) Conclusions

Examination of the minimal cut-set listing confirms that five sets comprise two rate terms, whilst the remainder contains a single rate term. All sets also contain probability terms. The analysis is therefore seen to comply with dimensional criteria requirements in that the fault-tree end event is a rate subscribed to by all the system cut sets which are each dimensionally seen to be that of rate.

The process system risk is seen to be most sensitive to a total of four second-order cut sets, namely *CE, CF, DE* and *DF*. Lower-order cut sets generally have the potential of providing the more significant risks and therefore should always be critically examined. At this stage of this demonstration study it is sufficient to say that second- and higher-order cut sets are also very pertinent to common-mode failure assessment which will be dealt with in Chapter 10.

Cut sets CE and CF
These two cut sets show that concurrent failures of either process shutdown valve V_A or V_B with that of the temperature controller TC will result in a process hazard. Process shutdown valves do not undergo frequent or continuous modulation and are also difficult to proof test on a continuous process application. For these reasons they are known to be inherently less reliable than safety input or logic interface components. Hence the failure state system pairs given by these two cut sets are seen to be significant in terms of overall plant system risk.

The temperature controller, if satisfactorily maintained, should present a relatively lower risk of failure. Hence the conclusion is made that the automatic shutdown system is a significantly weak area in terms of system reliability. This is attributable to lack of redundancy and facilities for more frequent proof testing.

Cut sets DE and DF
These two cut sets reveal that the elements of most significant risk are attributable to three automatic process valves, namely, the two shutdown valves and the modulating temperature control valve TCV. These automatic valves are inherently the most unreliable items in the overall system with the shutdown valves being the more significant. It is concluded that some redesign of the shutdown valve and modulating valve systems should be considered.

Cut sets ABGJ, ABHJ and ABHK
Each of these fourth-order cut sets relate to total failure of the cooling pump system and inherently reliable components of the safety system. Therefore the subsystems which the cut sets represent would not be

regarded as risk significant and hence not subject to modification. It should be noted that the combined failure rate of the two pumps would be relatively low compared with that of a single pump.

Cut sets ABE and ABF
These each represent a lower but not insignificant risk in that each is dependent on the low failure rate of the pump system when combined with the higher failure probability of either shutdown valve. The system hazard rate contribution from these two sets will therefore be less significant but not sufficiently low to be ignored. They suggest that the shutdown system is a significant item in terms of failure.

Cut sets DHJ, DHK, DGK and DGJ
All these sets refer to combinations of inherently reliable safety subsystems with the temperature control modulating valve TCV. The control valve, by virtue of its continuous modulating duty combined with improved maintainability when installed with a manually operated regulating bypass, would be expected to yield a more reliable process duty than that of a safety shutdown valve.

Cut sets CGJ and CHJ
These cut sets possibly provide the most insignificant contributions to overall system hazard rate. The temperature controller TC will be expected to exhibit a low failure rate in its dangerous mode. The combinations of highly reliable safety subsystems with the relatively reliable temperature controller should not give rise to any need of possible improvement.

(b) Recommendations

- The automatic shutdown system should be modified to include redundancy and facilities enabling more frequent on-line proof testing down to a minimum period of six weeks.
- The cooling pump system maintenance regime should be changed from breakdown to that of planned maintenance. In making this recommendation it is recognized that although the system will have a relatively low failure rate, further enhancement will be beneficial when meeting those risk conditions when shutdown valve failure states are also present.
- The modulating temperature control valve TCV should be installed with a manual bypass regulating valve facility. This would enable regular maintenance to be carried out without process interruption and hence minimize its contribution to overall process system hazard rate.

7.10.8 General observations on the demonstration study

The study has shown how qualitative assessment can result in valuable conclusions which offer enhancements in design and safety philosophies at early stages before commitment to significant plant capital expenditure. In initial process design the actual equipment hardware specifications are very often unknown and hence a specific data-based quantitative assessment is not always possible. An early qualitative assessment is therefore seen to be beneficial to project control in terms of meeting programme dates and overall economy.

The subject of this demonstration study has been chosen with respect to a relatively small and simplified process system for purposes of initial clarity and understanding. This has resulted in the derivation of a very small minimal cut-set listing and therefore it has been possible to base the conclusions and recommendations on a detailed consideration of each cut set.

In the practical case where systems are most often quite extensive the cut-set listing would contain a high number of set entries which would make detailed individual considerations abortive in terms of engineering time and effort.

In Chapter 8 it will be shown how cut-set listings are converted by factorization into a qualitative mathematical model which is a much simpler and convenient form from which to draw qualitative conclusions.

However, minimal cut-set listings are essential in their own right as a means of carrying out common-mode or dependent failure analyses and will be dealt with in some detail in Chapter 10.

8

Mathematical modelling

8.1 INTRODUCTION

The mathematical model represents the culmination of any safety and reliability assessment which seeks to establish the system level of risk against a given criterion. The model may be expressed either in qualitative or quantitative form, the choice being dictated by considerations of expediency in the early design stage and costs related to both process system capital outlay and assessment time available.

The qualitative model is recognized as a lower cost-sensitive option since it does not require data acquisition and associated difficulties in type matching and operational environmental factors. Its value to designers and operational managements lies in its ability to enable judicious decisions to be made at very early stages where designs can be more easily made without involving those cost penalties which would have been necessarily incurred when approaching the more finalized detailed process design stage.

Quantitative modelling relies on the application of representative data sets with the objective of concluding a value for system risk. The application of data sets necessitates accessibility to specific data, which is a first priority and ideally should be drawn from the user's particular experience. Assessment conclusions based on confidence of absolute values assigned to data is a desirable objective and can be attainable if the quality of data is present and is adequately applied. However, absolute values of data should generally be regarded with caution since a pessimistic overview can lead to unnecessary costs in redesign and equipment specifications as well as changes in operational management philosophy. An optimistic assessment, on the other hand, constitutes a most undesirable conclusion whereby the given process system could eventually operate in a more dangerous state than that set by the criterial requirement.

There is a third alternative to those of qualitative and absolute value assessments and which in many cases can be a very effective approach, particularly in the earlier project design stages. Quantified data-based modelling has an overriding merit in that weak and strong areas can be easily and rapidly identified with a greater degree of confidence than with that of the qualitative case. It follows that when design in principle is the consideration, then data set approximation can be a very powerful approach to optimization of the most favourable system design or re-design. A data set based on approximation when applied to the conceptual system will indicate weak and strong areas, but with unreliable absolute values. Repetitive design of weak areas using the same data set would again yield values which, in absolute terms, would not be regarded as reliable. However, comparisons of weak area designs into which the same data sets are applied will produce values which, in relative terms, are reliable indications of whether reliability or safety would be improved or degraded by the alternative design.

8.2 BASIS OF THE MATHEMATICAL MODEL

8.2.1 Fault-tree equivalence

Mathematical models are derived from engineering level fault trees which have been expressed in Boolean deterministic terms in the form of minimal cut-set listings. Where Boolean reduction has been necessary in order to produce a minimal cut-set listing, then factorizing of the minimal cut sets will produce a mathematical model which is equivalent to but often hardly recognizable in form to the parent fault tree. An engineering fault tree which yields only minimal cut sets without Boolean reduction would be an identical network to that of its derived mathematical model.

8.2.2 Progression

It was recommended in Chapter 7 that fault-tree progression should take place from left to right, commencing with the undesirable end event and terminating in the multiple primary failure events. Mathematical models which may be identical or non-identical in form to the parent fault tree are recommended to commence on the left-hand side with the primary failure events and terminating on the right with the undesired end event. The progression philosophy from left to right follows the common practice of reading.

8.3 EVALUATION OF THE MATHEMATICAL MODEL

8.3.1 The qualitative model

When a qualitative philosophy is being used in a risk assessment then treatments and conclusions drawn from logic gate combinations will be identical to those of the parent Boolean reduced fault tree.

8.3.2 The quantified model

When a quantification philosophy is being used the mathematical model, being probabilistic in nature, therefore requires classical algebraical treatment. This implies that failure events instead of having a universal discrete Boolean value of 1 will each instead have some algebraic value over the range of 0 and 1 corresponding to the chance or probability of it being present.

To meet a quantification requirement it is necessary to transpose from the Boolean deterministic viewpoint to that of the probabilistic. Such transpositions call for recognition of certain rules and procedures at all logic gates in order to derive a result which is commensurate with the aims of the deterministic model and its equivalent probabilistic expression, i.e. to establish a state of equivalence between the fault tree and the mathematical model.

8.4 QUANTITIES IN RISK ASSESSMENT MODELLING

Risk assessments are concerned with two quantities, namely probabilities and rates.

8.4.1 Probabilities

Probabilities represent a common denominator in all risk assessments. They occur wholly in protective system reliability studies and to a lesser extent in process plant demand assessments.

(a) Protective systems

A primary objective of any risk study is to assess in engineering terms the reliability performance of the plant protective system and is always carried out irrespective of any possible requirement to assess the frequencies of dangerous process plant demands. Performance is normally expressed in the form of a failure probability on demand, i.e. its mean fractional dead time and may refer to either of two changes of state, namely the partly reversible state and the reversible state. The former is pertinent to the safety function, i.e. dangerous unrevealed failures, whilst the latter is relevant to system unavailability in the safe or spurious tripping mode. In almost all safety assessments, performance in the partly reversible state is the predominant assessment requirement.

The protective system study in the safety mode, when quantified, would be expected to conclude with a statement of the system FDT, the acceptability of which would lie somewhere in the range of 10^{-1} to 10^{-5} depending on the nature of the process hazard and the plant demand rate. The range of quoted FDT values refers to the chance of protective system failure in the event of a process demand being presented to the input of the safety system. Hence 10^{-1} implies statistically that the protection will fail once in every ten process demands. It does not mean that for every ten demands the protection will always fail to respond once, but rather that over a very large number of demands the protective system failure will average once per ten demands. Similarly 10^{-5} implies that failure will occur on average once per 100 000 demands. The reliability rating therefore may be viewed as an engineering index which indicates a relatively higher or lower performance in terms of protective function.

It is outside the scope of this safety and reliability treatise to suggest relative protective system performance values which could apply to specific process hazard states. However, as a very rough guide, higher system FDT values, say 10^{-2} to 10^{-4}, may be acceptable for the more minor hazard scenarios or where process demand rates are relatively low. For more major hazards coupled with less frequent or more frequent demands, acceptable FDT values may lie in the range of 10^{-4} to 10^{-5}. The former would be associated with established standards of engineering practice, whereas the latter would conform to higher traditional 'Rolls Royce' standards with associated higher costings in terms of capital outlay, testing and maintenance.

Protective system performances in terms of safety function are judged solely in probability terms in accordance with the mathematics of the partly reversible state. It follows that the function of each of the logic gates used in the mathematical modelling of such systems is to process a probability input element set to a single FDT probability output.

(b) Process plant systems

Probabilities are widely prominent in assessments of process plant demands. When assessing a specific demand, i.e. frequency of a defined process abnormality which could lead to a hazard, probabilities, if present, will always be combined with rate quantities in order to express the specific demand.

In a process demand evaluation, probabilities may refer to either one or both changes of state, namely, the partly reversible state and/or the reversible state which have already been discussed in a previous chapter.

In practice, process demands are revealed by appropriate sensors which compose the input subsystem of the related protective system. When probability factors are present in a demand logic they are assessed in terms of mean fractional dead times based on the appropriate change of state mode.

8.4.2 Rates

In risk assessments, rates are pertinent solely to evaluations of dangerous process demand frequencies. They are only present in protective system assessments in the reversible state or spurious tripping mode and are therefore acknowledged as safe failures.

(a) Protective systems

On certain hazardous process plants, enforced shutdowns due to safety system malfunctions may result in significant monetary penalties as a result of loss of production. Additionally, higher risk scenarios may be present during the period of start up on certain hazardous processes. When these considerations are regarded as being significant, the annual spurious tripping rate of the protective system may need to be taken into account and its acceptability verified.

Protective system actions in the absence of plant demands and which result in plant process shutdowns are classified as revealed failures. The incidences of such failures are evaluated in accordance with the principles discussed for the reversible state in Chapter 3, with particular reference to effective system failure rate.

(b) Process plant systems

A process plant risk study involves assessments of annual frequencies of dangerous demands and hence rates are predominant quantities in the evaluation of plant risk. A dangerous demand is constituted by a process abnormality which arises because of failures in the process system. These failures may be due to equipment faults, human errors in operation and maintenance or management software. The related protective system employs specific sensors which monitor vital plant conditions in order to initiate process shutdowns or operator alarms when the particular condition goes outside the designated safe limit.

Demand frequency fundamentally comprises rate element sets which may be serial or parallel in nature. The assembly of these rate elements will produce an overall rate quantity which describes the dangerous plant demand. All faults, whether equipment or human error based, must be expressed in terms of dangerous failure modes applicable to the process application. If the elements of the demand are serial then the overall frequency is simply the sum of the dangerous failure rates.

Parallel elements comprising, say pump systems, heaters, valve systems etc. will include probability factors necessary to achieve a statement of overall subsystem failure rate. An example of such a parallel system would be illustrated by, say a hypothetical pumping system consisting of two pumps arranged such that either pump will provide the necessary process duty. Loss of both pumps would constitute a dangerous demand and therefore probability factors would feature in the pump system failure rate evaluation. Hence for process plant demand mechanisms which contain parallel elements, both rates and probabilities will be necessary features in an assessment.

The following important conclusions may therefore be stated in respect of the treatment of multiple rate element sets in the assessment of process demands.

- Serial elements are assembled through OR gate logics in order to derive the representative overall demand rate.
- Parallel elements accompanied by probability factors are assembled through AND gate logics in order to derive the representative overall demand rate.

8.5 LOGIC GATES IN SYSTEMS SAFETY MODELLING

8.5.1 Logic gate types

Risk assessment mathematical modelling may be most conveniently confined to the use of three types of gate:

- OR gates
- AND gates
- majority voting gates.

8.5.2 Dimensional criteria at logic gates

The derived mathematical model must firstly comply at each of its logic gates with dimensional criteria already discussed in Chapter 3. It follows that if the parent fault tree complies fully with dimensional criteria, and reduction processes have been correctly carried out in accordance with Boolean relationships, then the ensuing mathematical model would likewise conform to criterial requirements at each of the logic gates in the engineering fault-tree model.

8.5.3 Legal and illegal inputs

With the exception of the OR gate, mixed inputs of rates and probabilities are legal. In the case of the OR gate, mixed inputs are illegal since the output would not constitute a pure rate or probability and hence does not comply with dimensional criterial requirements.

8.5.4 Serial and parallel input elements

If a system failure is dependent on a number of elements such that failure of any one would bring about system failure, then such elements are said to be serial and would be processed logically through an OR gate. Conversely if system failure is dependent on coincident failure of all elements, then such elements are said to be parallel and would be processed logically through an AND gate.

8.6 BOOLEAN COMBINATIONS AT LOGIC GATES

In practice, safety assessments are carried out by the use of OR gates, AND gates and majority voting gates. The employment of NOT logics in the form of NAND (not AND) and NOR (not OR) is an unnecessary refinement which complicates evaluations and does little to enhance the overall confidence of the ultimate safety assessment when viewed against quality levels of statistical data sets. In presenting the methodology of practical mathematical modelling, the use of NOT logics will therefore be omitted from consideration.

The minimum number of input elements to the three types of logic gates to be considered will always be two, with no set upper limit. In mathematical modelling it is found that the number of inputs is most likely to encompass two, three or four, and very occasionally up to five.

The resolution of inputs to the single output at any given logic gate may be most conveniently carried out by means of a truth table which identifies all the possible relevant failure states in a Boolean deterministic format. The overall gate logic output is given by the sum of the designated failure states in the table which later enables transposition into the equivalent probabilistic expression demanded by the requirements of the particular assessment.

The probabilistic evaluation methodology to be described is applicable to any number of inputs and therefore, in order to conveniently illustrate the evaluation techniques, a gate system reference standard of three inputs will be considered for each of the three types of logic.

Table 8.1 sets out the truth table for three elements A, B and C and the related failure states in respect of the three logic gate types under consideration. For a given element in the table:

- A symbolizes that A is in a failed state;
- $\bar{A}$ (not A) symbolizes that a failure state is not present in A, i.e. the element is in a success state.

Table 8.1 Truth table – three element input logic gates

State	Elements			Failure states		
	A	B	C	OR	AND	2/3
1	$\bar{A}$	$\bar{B}$	$\bar{C}$			
2	$\bar{A}$	$\bar{B}$	C	✓		
3	$\bar{A}$	B	$\bar{C}$	✓		
4	$\bar{A}$	B	C	✓		✓
5	A	$\bar{B}$	$\bar{C}$	✓		
6	A	$\bar{B}$	C	✓		✓
7	A	B	$\bar{C}$	✓		✓
8	A	B	C	✓	✓	✓

The Boolean statements to be derived for each of the three logic gates are each complete in their respective entireties in that they do not require any further interpretation when applied to the varying requirements of mathematical modelling. They represent valid expressions for separate probability and rate input sets and combinations of both.

Boolean logic system expressions are necessary precursors to mathematical modelling in that they provide the basis for all the probabilistic determinations in the eventual model.

Referring to Table 8.1 and using the Boolean relationships defined in section 6.2, expressions for the three gate system types shown in the table will now be derived in preparation for their use in the probabilistic stage to be described later.

8.6.1 Three-input OR gate

Table 8.1 shows that system failure given by the output of the OR gate comprises the group of failure states 2–8 inclusive. Hence (noting that Boolean OR is signified by '+')

System failure = Logic gate output

$$= \text{State 2 OR State 3 OR} \ldots \text{State 8}$$

$$= \bar{A}\bar{B}C + \bar{A}B\bar{C} + \bar{A}BC + A\bar{B}\bar{C} + A\bar{B}C + AB\bar{C} + ABC$$

$$= \bar{A}\bar{B}C + \bar{A}B(\bar{C} + C) + A\bar{B}(\bar{C} + C) + AB(\bar{C} + C)$$

$$= \bar{A}\bar{B}C + \bar{A}B + A\bar{B} + AB \quad (\text{since } \bar{C} + C = 1)$$

$$= \bar{A}\bar{B}C + \bar{A}B + A(\bar{B} + B)$$

$$= \bar{A}\bar{B}C + \bar{A}B + A$$

$$= \bar{A}(\bar{B}C + B) + A$$

$$= \bar{A}(C + B) + A \quad (\text{since } \bar{B}C + B = C + B)$$

$$= \bar{A}C + \bar{A}B + A$$

$$= \bar{A}C + B + A \quad (\text{since } \bar{A}B + A = B + A)$$

$$= C + A + B \quad (\text{since } \bar{A}C + A = C + A)$$

Hence

Boolean output from three-input OR gate failure logic $= A + B + C$

8.6.2 Three-input AND gate

From Table 8.1, system failure is given by state 8. Hence:

Boolean output from three-input AND gate failure logic $= ABC$

8.6.3 Three-input 2/3 majority voting gate

Table 8.1 shows that states 4, 6, 7 and 8 are the relevant failure states for the 2/3 majority voting failure logic. Hence

System failure = State 4 OR State 6 OR State 7 OR State 8

$$= \bar{A}BC + A\bar{B}C + AB\bar{C} + ABC$$

$$= \bar{A}BC + A\bar{B}C + AB(\bar{C} + C)$$

$$= \bar{A}BC + A\bar{B}C + AB$$

$$= \bar{A}BC + A(\bar{B}C + B)$$

$$= \bar{A}BC + A(C + B)$$

$$= \bar{A}BC + AC + AB$$

$$= C(\bar{A}B + A) + AB$$

$$= C(B + A) + AB$$

$$= CB + CA + AB$$

Hence:

> Boolean output for three-input majority voting failure logic
>
> $$= AB + AC + BC$$

8.6.4 Summary – Boolean gate logics for three inputs

The three failure logic gates used in mathematical modelling are now presented in Figure 8.1 with relevant Boolean output identifying equations in respect of three input elements A, B and C.

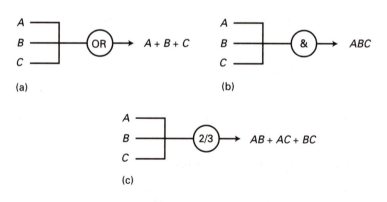

Figure 8.1 The three failure logic gates.

8.7 PROBABILISTIC – DEFINITION

Probabilistic expressions refer to both probabilities and rates and are non-Boolean in form. A mathematical model is a probabilistic expression which affords both qualitative and quantitative conclusions regarding levels of risk in terms of probability or frequency of failure.

Unlike Boolean logic, related probabilistic expressions at logic gates may be subject to different evaluations dependent on the following:

- magnitude of individual probability inputs at OR gates;
- number of probability inputs at OR gates;
- output requirement in terms of maximum failure probability;
- output requirement in terms of mean fractional dead time;
- input and output requirements in terms of rates at AND gates.

8.8 PROBABILITY EXPRESSION NOMENCLATURES

A probability may be expressed in either of two modes, namely maximum failure probability or mean fractional dead time. For convenience:

- **maximum failure probability** will be referred to as probability P;
- **mean fractional dead time** or average failure probability on demand will be referred to as FDT (μ).

8.9 DEFINITIONS OF PROBABILITY – PARTLY REVERSIBLE STATE

- **Failure probability** will refer to the value of the exponential failure probability at the end of the proof test interval.
- **Mean fractional dead time** will refer to the average value of the exponential failure probability over the proof test interval, i.e.

$$\mu = \frac{1}{\tau} \int_0^{\tau} P \, dt$$

where τ is the proof test interval.

8.10 DEFINITIONS OF PROBABILITY – REVERSIBLE STATE

- **Failure probability** P will refer to the time-independent value of probability over the mean time between failures given by $\theta_r \tau_r$, where θ_r is the revealed or spurious failure rate and τ_r is the mean repair time.
- **Mean fractional dead time** μ will refer to the constant failure probability over the mean time between failures, this again being given by $\theta_r \tau_r$.

Hence

$$P = \mu.$$

8.11 HIGHER-ORDER PROBABILITY TERMS IN GATE OUTPUTS

Irrespective of failure state mode, probability sets taken through OR and majority voting gates always contain higher-order terms in their output expressions which are usually neglected because failure probabilities are normally low.

When single-order probability elements in an OR gate input are such that their simple addition yields a probability output which approaches, or is in excess of, unity, then the use of higher-order output terms must be taken into account. This also applies to majority voting gates when second-order terms in the output approach unity.

8.12 COMBINING HIGHER-ORDER PROBABILITIES

8.12.1 Partly reversible state

The method of evaluation of higher-order probability terms in the partly reversible state is subject to the presence or absence of mean fractional dead times associated with the higher-order terms. Where OR gates are present, evaluation of higher-order terms in a gate output may be necessary in order to achieve an acceptable level of quantification in accordance with conditions stated above in section 8.7. Whenever a higher-order probability term, such as P^2, occurs, then the derived FDT, which is a measure of average probability over a given period of time τ, is given by

$$\mu = \frac{1}{\tau} \int_0^\tau P^2 \, dt$$

Consider a group of, say three identical FDTs which are to be multiplied in order to derive a representative FDT output. Let the associated failure probability P of each of the input elements be equal to $\theta\tau$, where θ is the unrevealed failure rate and τ is the proof test interval. Then by integration,

$$\text{Single element FDT} = \frac{\theta\tau}{2}$$

Multiplying the three input FDTs gives

$$\text{Gate output FDT} = \frac{\theta^3\tau^3}{8} \tag{8.1}$$

However, if the system is initially expressed as a maximum failure probability of P^3, then

$$\text{Overall system FDT} = \frac{1}{\tau} \int_0^\tau P^3 \, dt$$

$$= \frac{1}{\tau} \int_0^\tau \theta^3 t^3 \, dt$$

$$= \frac{\theta^3 \tau^3}{4} \tag{8.2}$$

Comparing equations 8.1 and 8.2 shows that the former is optimistically in error and constitutes a dangerous conclusion from a safety point of view.

It is therefore concluded that fractional dead times derived from higher-order terms in the partly reversible state failure mode are determined through the process of integration in order to derive the single equivalent FDT.

8.12.2 Reversible state

In the reversible state mode, probabilities are also mean fractional dead times and therefore when higher-order terms are present they are simply evaluated by taking the product of the factors in each of the specific terms.

8.13 RATE COMBINATIONS AT THE LOGICAL OR GATE

OR gate failure logic is relevant to failures of serial elements in that failure of any one element will bring about overall system failure. The identifying Boolean expression for the three-input OR gate is given as $A + B + C$ and is shown with relevant logic representation in Figure 8.1(a). The expression is applicable to either rates or probabilities but not a combination of both.

Table 8.1 and section 8.6.1 show that for a three-input OR gate system consisting of elements A, B and C, the output in Boolean terms is $A + B + C$ (where '+' means OR).

In probabilistic terms, let:

- the rate quantity synonymous with input element A be R_A;
- the gate output system rate be R_0.

    Then

$$R_0 = R_A + R_B + R_C$$

Hence rates are combined at OR gates simply by addition of the input terms with no restriction on the magnitude of the resultant output value, i.e. R_0 may be >1.0.

8.13.1 Partly reversible state

Failures in this state are unrevealed; therefore in general terms R would refer to a dangerous failure state mode and would have a particular relevance to system safety.

8.13.2 Reversible state

Failures in this state are revealed and are regarded in a safety sense as safe. Therefore in general terms R would refer to a safe failure mode and would have relevance to system availability.

8.13.3 Logical representation for rates at the OR gate

In both the partly reversible and reversible state failure modes the probability logic is identical and may be defined as in Figure 8.2, subject to appropriate unrevealed and revealed failure classifications of input element sets.

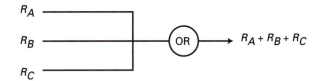

Figure 8.2 OR gate rate logic ('+' refers to addition).

8.14 PROBABILITY COMBINATIONS AT THE LOGICAL OR GATE

Probability transitions through OR gates are generally simple to apply providing the individual input element probability magnitudes are reasonably small, i.e. <0.1, or fewer in number. In these cases, which account for a very large majority, the gate inputs need only be added together in order to express the output value. However, even though the input probability values may be quite low, there are instances where there may be a large number of elements in the input set which, if added together, could result in an output value equal to or greater than unity. In these cases the assessor will need to apply the overall probability equation which will contain higher-order terms. These terms would be taken into account until a satisfactory terminating value was achieved for the overall system probability value.

Referring to the truth table, Table 8.1, an overall system failure probability expression for a three-input OR gate having elements A, B and C, is derived as follows.

- Let P_A be the failure probability associated with element A etc.
- Let $(1 - P_A)$ be the success probability associated with element $\bar{A}$ etc.
- Let P_0 be the gate output or system failure probability.

Initially for simplification let

$$P_A = P_B = P_C = P$$

Then by example,

$$\text{State 2 in Table 8.1} = P(1 - P)^2$$

Hence by summing states 2–8 inclusive for the OR gate,

$$P_0 = 3P(1 - P)^2 + 3P^2(1 - P) + P^3$$

$$= 3P - 6P^2 + 3P^3 + 3P^2 - 3P^3 + P^3$$

$$= 3P - 3P^2 + P^3 \tag{8.3}$$

Substituting for P_A, P_B and P_C into equation 8.3 gives:

$$P_0 = (P_A + P_B + P_C) - (P_A P_B + P_A P_C + P_B P_C) + (P_A P_B P_C) \tag{8.4}$$

Constructing truth tables in excess of three inputs is time consuming and therefore a more convenient direct approach would be appropriate. This can be achieved by means of Pascal's triangle suitably formulated for the given task. From the triangle, coefficients of the probability function can be directly read off and the composite probability equation conveniently expressed. Figure 8.3 shows Pascal's triangle with a redundant section removed and compiled for a maximum of five probability input elements.

To demonstrate the use of the triangle refer to the '3' input reference which gives the coefficient '3' for the first term, and '3' for the second term, and '1' for the last term. Hence the terms are

$$3P - 3P^2 + P^3$$

The probability OR gate will now be considered in the various formats demanded by differing safety assessments. In each of the descriptions, for purposes of simplification it may be initially assumed that

$$P_A = P_B = P_C = P$$

Formatting of gate outputs will be carried out with respect to

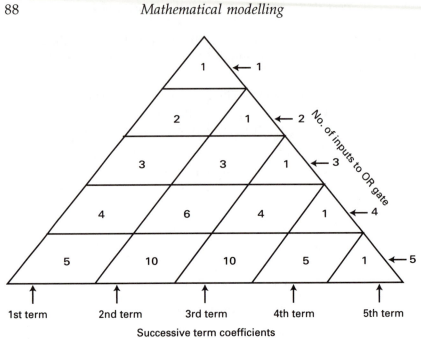

Figure 8.3 Pascal's triangle for OR gate – one to five probability inputs.

- the partly reversible failure state;
- the reversible failure state.

Let

$$P = \text{failure probability}$$
$$\theta = \text{unrevealed failure rate}$$
$$\theta_r = \text{revealed failure rate}$$
$$\tau = \text{proof test interval}$$
$$\tau_r = \text{mean repair time}$$
$$\mu = \text{mean fractional dead time FDT}$$
$$t = \text{time}$$

where

$$P = \theta\tau \quad \text{(unrevealed)}$$
$$P = \theta_r\tau_r \quad \text{(revealed)}$$

8.14.1 Partly reversible state

In this mode failure rates are unrevealed and are subject to periodic proof-testing procedures.

(a) Probability inputs → probability output

$$\text{Inputs } = P_A + P_B + P_C$$

$$\text{Output} = P_0 = 3P - 3P^2 + P^3$$

Higher-order terms neglected

$$\text{Output} = 3P = \theta_A \tau + \theta_B \tau + \theta_C \tau$$

$$= (\theta_A + \theta_B + \theta_C)\tau$$

Third-order term only neglected (subject to acceptable error margin)

$$\text{Output} = (P_A + P_B + P_C) - (P_A P_B + P_A P_C + P_B P_C)$$

$$= (\theta_A + \theta_B + \theta_C)\tau - (\theta_A \theta_B + \theta_A \theta_C + \theta_B \theta_C)\tau^2$$

(b) Probability inputs → FDT output

$$\text{Inputs } = P_A + P_B + P_C$$

$$\text{Output} = \mu_0$$

Higher-order terms neglected

$$\text{Output} = \mu_0 = \frac{1}{\tau} \int_0^\tau 3P \, dt = \frac{3\theta\tau}{2}$$

i.e.

$$\mu_0 = \frac{\tau}{2}(\theta_A + \theta_B + \theta_C)$$

Third-order term only neglected

$$\text{Output} = \mu_0 = \frac{1}{\tau} \int_0^\tau 3P - 3P^2 \, dt$$

$$\mu_0 = \frac{1}{\tau} \int_0^\tau 3\theta t - 3\theta^2 t^2 \, dt$$

$$\mu_0 = \frac{3\theta\tau}{2} - \frac{3\theta^2\tau^2}{3}$$

$$= \frac{\tau}{2}(\theta_A + \theta_B + \theta_C) - \frac{\tau^2}{3}(\theta_A \theta_B + \theta_A \theta_C + \theta_B \theta_C)$$

(c) FDT inputs → FDT output

$$\text{Inputs} = \mu_A + \mu_B + \mu_C$$

$$\text{Output} = \mu_0$$

Convert input FDTs to equivalent probabilities, i.e.

$$P_A = 2\mu_A \text{ etc.}$$

The output is expressed in accordance with section 8.14.1(b).

8.14.2 Reversible state

The reversible state refers to revealed failure rates and probabilities for systems and subsystems which are significant to repair times. Failure probability in this state is independent of time and hence it is also the mean fractional dead time.

Failure probability P or FDT μ in the reversible state has previously been shown to be

$$P = \mu = \frac{\theta_r \tau_r}{1 + \theta_r \tau_r} = \theta_r \tau_r \text{ (when } \theta_r \tau_r < 0.1)$$

The various formats demanded by differing assessments will now be considered.

(a) Probability inputs → probability output

$$\text{Inputs} = P_A + P_B + P_C = \mu_A + \mu_B + \mu_C$$

$$\text{Output} = P_0 = \mu_0$$

Higher-order terms neglected

$$P_0 = \mu_0 = 3P = 3\mu$$

$$= \theta_A \tau_r + \theta_B \tau_r + \theta_C \tau_r$$

$$= \tau_r(\theta_A + \theta_B + \theta_C)$$

Third-order term only neglected

$$P_0 = \mu_0 = 3P - 3P^2$$

$$\text{Output} = \tau_r(\theta_A + \theta_B + \theta_C) - (\theta_A \theta_B + \theta_A \theta_C + \theta_B \theta_C)(\tau_r)^2$$

8.15 RATE COMBINATIONS AT THE LOGICAL AND GATE

The identifying Boolean expression for the three-input AND gate is given in logical form in Figure 8.1(b) as *ABC*. AND gate failure logic is relevant to failures of parallel elements such that all elements must fail in order to produce a gate output representative of system failure. Boolean logic indicates that quantities are to be multiplied in transit across the AND gate. In expressing the probabilistic equivalent to the Boolean directive, the guiding principle must be such that dimensional criteria are not violated.

The direct product of the input elements of a rate set will produce a higher-order gate output result of the form R^n, where n is the number of elements in the set. Clearly, direct multiplication of rate elements will violate dimensional criteria in that the gate output would result in T^{-n} instead of the required rate dimension of T^{-1}. The representative gate logic for the three-input system *A*, *B* and *C* in terms of rate may be initially defined as in Figure 8.4.

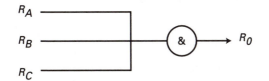

Figure 8.4 AND gate rate logic.

In order to derive a probabilistic expression from multiple rate inputs R_A, R_B and R_C at the AND gate it is expedient to use a truth table approach. Hence referring to the logic gate system shown in Figure 8.4: Let R_A be the failure rate of input element *A* etc., R_0 be the gate output or system failure rate and μ_A be the probability on demand of *A* being failed etc. Hence R_0 is derived as follows:

State 1 = (Rate of failure of *A*) × (FDT of *B* failure) × (FDT of *C* failure)

State 2 = (Rate of failure of *B*) × (FDT of *A* failure) × (FDT of *C* failure)

State 3 = (Rate of failure of *C*) × (FDT of *A* failure) × (FDT of *B* failure)

Therefore

System failure rate R_0 = Sum of states 1, 2 and 3

i.e.

$$R_0 = R_A \mu_B \mu_C + R_B \mu_A \mu_C + R_C \mu_A \mu_B \tag{8.5}$$

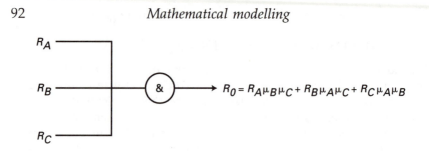

Figure 8.5 Derived expression of Figure 8.4.

Hence the logic in Figure 8.4 can now be expressed generally as in Figure 8.5 for both the partly reversible and reversible failure modes.

In a quantified model it would be usual for rates to be expressed in terms of faults per year and failure probability factors μ as mean fractional dead times. The rate and probability pairs must each be referred to the pertinent operational failure state mode, namely the partly reversible or the reversible, synonymous respectively with the unrevealed or revealed modes of failure rate.

8.15.1 Partly reversible state

This refers to unrevealed failures in a system of parallel rate elements, which are associated with process demand mechanisms. The identifying general equation is given as equation = 8.5 above whose probability factors will be assessed on the basis of mean fractional dead times in accordance with the mathematics of the partly reversible state. In this state, if $P < 0.1$, then

$$P = R\tau$$

where R is the unrevealed failure rate and τ is the process maintenance regime interval. Then the FDT is

$$\mu = \frac{1}{\tau} \int_0^\tau P\,dt = \frac{R\tau}{2}$$

When associated with process demands, τ is relevant to an inspection and maintenance regime which, on a process plant for example, may be annually or bi-annually at planned process shutdown. If annually,

$$\tau = 1.0 \quad \text{and} \quad \text{FDT } \mu = R/2$$

If bi-annually,

$$\tau = 0.5 \quad \text{and} \quad \text{FDT } \mu = R/4$$

Hence for the three-element AND gate shown in Figure 8.5 on a bi-annual basis the general output rate equation 8.5 would be expressed as follows:

$$R_0 = \frac{R_A R_B R_C}{12} + \frac{R_B R_A R_C}{12} + \frac{R_C R_A R_B}{12} \qquad (8.6)$$

Note on the FDT factors $R_B R_C / 12$ etc.

Referring to the term $R_A \mu_B \mu_C$ in equation 8.5

$$\mu_B = \text{FDT associated with rate element } B \text{ etc.} = \frac{R_B \tau}{2}$$

Therefore

$$P_B = R_B \tau$$

Hence $\mu_B \mu_C$ is derived from

$$\frac{1}{\tau} \int_0^\tau P_B P_C \, dt$$

i.e.

$$\mu_B \mu_C = \frac{1}{\tau} \int_0^\tau R_B R_C t^2 \, dt$$

$$= \frac{R_B R_C \tau^2}{3}$$

Let $\tau = 0.5$ (bi-annual inspection), then

$$\mu_B \mu_C = \frac{R_B R_C}{12}$$

Also if $\tau = 1$ (annual inspection), then

$$\mu_B \mu_C = \frac{R_B R_C}{3} \text{ etc.}$$

8.15.2 Reversible state

This refers to revealed failures in a system of parallel elements which are associated with process demand mechanisms. The identifying equation is again that of equation 8.5 above, the probability factors of which are assessed on the basis of mean fractional dead times in accordance with the mathematics of the reversible state. In this state, if $P < 0.1$, then

$$P = R \tau_r$$

where R is the revealed failure rate and τ_r is the mean repair time. Then the FDT is

$$\mu = P = R\tau_r$$

Hence for the three-element AND gate system shown in Figure 8.5,

$$R_0 = R_A R_B R_C (\tau_r)^2 + R_B R_A R_C (\tau_r)^2 + R_C R_A R_B (\tau_r)^2$$

Let $\tau = 0.5$ (bi-annual inspection), then

$$\mu_B \mu_C = \frac{R_B R_C}{4} \text{ etc.}$$

Also if $\tau = 1.0$ (annual inspection), then

$$\mu_B \mu_C = R_B R_C \text{ etc.}$$

8.16 PROBABILITY COMBINATIONS AT THE LOGICAL AND GATE

The relevant AND logic gate is represented as follows by Figure 8.6 and is given in terms of probabilities. Table 8.1 shows that state 8 is the only state which results in system failure, i.e. produces a gate output.

Let P_A be the failure probability associated with element A etc., $(1 - P_A)$ be the success probability associated with $\bar{A}$ etc. and P_0 be the gate output or system failure probability. Hence from state 8 in Table 8.1,

$$P_0 = P_A P_B P_C$$

The probability AND gate will now be considered in the various formats demanded by differing safety assessments. In each of the descriptions, for purposes of simplification, it may be initially assumed that

$$P_A = P_B = P_C = P$$

Formatting of gate outputs will be carried out with respect to

- the partly reversible failure state;
- the reversible failure state.

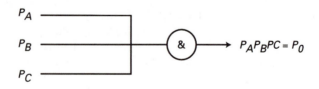

Figure 8.6 AND gate probability logic.

8.16.1 Partly reversible state

In this state failures are unrevealed and equipments are subject to periodic proof test. Let

$$P = \theta\tau$$

where θ is the unrevealed failure rate and τ is the proof test interval.

(a) *Probability inputs → probability output*

$$\text{Inputs} = P_A P_B P_C$$

$$\text{Output} = P_0 = P_A P_B P_C = \theta_A \theta_B \theta_C \tau^3$$

(b) *Probability inputs → FDT output*

$$\text{Inputs} = P_A P_B P_C$$

$$\text{Output} = \mu_0$$

$$\mu_0 = \frac{1}{\tau} \int_0^\tau P_A P_B P_C \, dt$$

$$\frac{1}{\tau} \int_0^\tau \theta_A \theta_B \theta_C t^3 \, dt$$

$$\mu_0 = \frac{\theta_A \theta_B \theta_C \tau^3}{4}$$

(c) *FDT inputs → FDT output*

$$\text{Inputs} = \mu_A, \ \mu_B, \ \mu_C = \frac{\theta_A \tau}{2}, \ \frac{\theta_B \tau}{2}, \ \frac{\theta_C \tau}{2}$$

Hence the inputs in terms of P_A etc. are

$$P_A, \ P_B, \ P_C = \theta_A \tau, \ \theta_B \tau, \ \theta_C \tau$$

Therefore the output FDT is

$$\mu_0 = \frac{1}{\tau} \int_0^\tau \theta_A \theta_B \theta_C t^3 \, dt$$

$$\mu_0 = \frac{\theta_A \theta_B \theta_C \tau^3}{4}$$

8.16.2 Reversible state

Owing to the independence of the input probability elements with time, failure probabilities in the reversible state when combined at AND gates are directly multiplied. Hence with reference to the three-element AND gate system shown in Figure 8.6, let

$$P = \mu = \theta_r \tau_r$$

where θ_r is the revealed failure rate and τ_r is the mean repair time.

$$\text{Output} = P_A P_B P_C = \mu_A \mu_B \mu_C$$

Hence

$$P_0 = \mu_0 = \theta_A \theta_B \theta_C (\tau_r)^3$$

8.17 RATE AND PROBABILITY INPUTS AT THE LOGICAL AND GATE

In dimensional terms, rates $[T^{-1}]$ are combined with probabilities [1] to provide system rates. Combined inputs are normally associated with risk assessments whereby demands are combined with protective system failure probabilities in order to yield process system hazard rates. The Boolean expression for a three-input AND gate given in Figure 8.1 as ABC may be applied to both rates and probabilities such that at least one input will be either a rate or a probability.

The mixed inputs will now be considered in the various formats demanded by differing assessments.

8.17.1 Partly reversible state

In this mode all failures are unrevealed and equipment is subject to proof testing regimes. Probability FDT factors associated with process subsystem breakdown rates refer to periodic process maintenance inspections.

Probabilities associated with protective equipment failure rates refer to proof testing intervals associated with safety systems.

Let

$$R_0 = \text{gate output rate}$$

$$R = \text{unrevealed failure rate of process equipment}$$

$$\tau_m = \text{process equipment inspection interval}$$

$$\theta = \text{unrevealed failure rate of protective equipment}$$

τ = proof test interval related to protective equipment

μ = FDT

t = time

Refer to Figure 8.1(b) for the AND gate.

(a) *Inputs (rates A and B, probability C) → output R_0*

$$\text{Output } R_0 = \left(\frac{R_A R_B \tau_m}{2} + \frac{R_B R_A \tau_m}{2}\right)\theta_C \tau$$

(b) *Inputs (rate A, probabilities B and C) → output R_0*

$$\text{Output } R_0 = R_A \theta_B \theta_C \tau^2$$

(c) *Inputs (rate A, FDTs B and C) → output R_0*

$$\text{Output } R_0 = \frac{R_A}{\tau}\int_0^\tau \theta_B \theta_C t^2 \, dt$$

$$R_0 = \frac{R_A \theta_B \theta_C \tau^2}{3}$$

8.17.2 Reversible state

In this mode all failures are revealed and equipments are subject to mean repair times.
   Let

R_0 = gate output rate

R = revealed failure rate of process equipment

τ_m = process equipment inspection interval

θ = revealed failure rate of protective equipment

τ_r = mean repair time related to protective equipment

μ = FDT

(a) *Inputs (rates A and B, probability C) → output R_0*

$$\text{Output } R_0 = (R_A R_B \tau_m + R_B R_A \tau_m)\theta_C \tau_r$$

(b) Inputs (rate A, probabilities B and C) → output R_0

$$\text{Output } R_0 = R_A \theta_B \theta_C (\tau_r)^2$$

(c) Inputs (rate A, FDTs B and C) → output R_0

$$\text{Output} = R_A \theta_B \theta_C (\tau_r)^2$$

Note that the output is also the expression derived under section 8.17.1(b) above since $P_B = \mu_B$ in the revealed state.

8.18 RATE COMBINATIONS AT MAJORITY VOTING GATES

In risk assessment modelling, majority voting gates provide two functions, namely processing of probabilities and processing of event frequencies or rates. The modelling principles to be described will be conveniently illustrated by consideration of a two-out-of-three voting failure logic. The principles to be illustrated are equally applicable to other voting logic systems.

The identifying Boolean expression for two-out-of-three failure voting logic consisting of three inputs *A*, *B* and *C* is given by Figure 8.1(c) as $AB + AC + BC$.

Modelling rates through majority voting gates are confined to revealed failures synonymous with the reversible state described in Chapter 3. They are concerned with failures of process plant equipments such as pumps, heaters or valves etc. which may lead to hazardous situations. Revealed failure rates are also pertinent to spurious operations of protective systems which may incur severe monetary losses due to unscheduled plant shutdowns.

Process plant rate inputs in majority voting configurations in the reversible state can constitute safe or dangerous conditions such as:

- safe process plant shutdowns due to loss of redundant process plant subsystems;
- dangerous process plant demands due to loss of those redundant systems which may lead to process hazard states;
- spurious protective system operations in the absence of dangerous plant demands, which result in costly plant shutdowns.

8.18.1 General equation – revealed rates in voting logic

Overall redundant system failure rates in the reversible state are given by the general equation for system spurious trip rate under section 3.3.3(a) which is reproduced here for convenience:

$$\theta_e = \left|\begin{matrix} n \\ r \end{matrix}\right| r\theta(\theta\tau_r)^{m-1}$$

where

θ_e = system effective spurious trip rate

θ = single subsystem apparent spurious trip rate

τ_r = mean repair time of single subsystem

n = minimum number of subsystems required for system success

r = minimum number of subsystems required to give system failure

such that

$$r = n - m + 1$$

8.19 PROBABILITY COMBINATIONS AT MAJORITY VOTING GATES

The truth table given as Table 8.1 shows that failure states 4, 6, 7 and 8 provide the gate logic output failure expression which is now to be converted to the equivalent probabilistic form.

Let P_A be the failure probability associated with element A etc., $(1 - P_A)$ be the success probability associated with element $\bar{A}$ and P_0 be the gate output or system failure probability. Initially for convenience let $P_A = P_B = P_C = P$. Then

$$P_0 = \text{(state 4)} + \text{(state 6)} + \text{(state 7)} + \text{(state 8)}$$

$$= P^2(1 - P) + P^2(1 - P) + P^2(1 - P) + P^3$$

$$= 3P^2(1 - P) + P^3$$

$$= 3P^2 - 3P^3 + P^3$$

$$= 3P^2 - 2P^3$$

Hence in terms of A, B and C

$$P_0 = P_A P_B + P_A P_C + P_B P_C - 2P_A P_B P_C \tag{8.7}$$

The probability 2/3 majority voting logic will now be considered in the various formats demanded by differing safety assessments.

8.19.1 Partly reversible state

In this mode all failures are unrevealed and equipment is subject to periodic proof testing. Let

$$\theta = \text{unrevealed failure rate}$$
$$\tau = \text{proof test interval}$$
$$\mu = \text{mean fractional dead time FDT}$$
$$t = \text{time}$$
$$P = \text{failure probability} = \theta\tau$$

(a) Probability inputs → probability output

$$\text{Inputs} = P_A, P_B, P_C$$
$$\text{Output} = P_0$$
$$= P_A P_B + P_A P_C + P_B P_C - 2P_A P_B P_C \quad \text{(equation 8.7)}$$
$$= \theta_A \theta_B \tau^2 + \theta_A \theta_C \tau^2 + \theta_B \theta_C \tau^2 - 2\theta_A \theta_B \theta_C \tau^3$$
$$= \tau^2(\theta_A \theta_B + \theta_A \theta_C + \theta_B \theta_C) - 2\tau^3(\theta_A \theta_B \theta_C)$$

(b) Probability inputs → FDT output

$$\text{Inputs} = P_A, P_B, P_C$$
$$\text{Output} = \mu_0$$

$$\mu_0 = \frac{1}{\tau}\int_0^\tau P_0 \, dt$$

$$= \frac{\theta_A \theta_B \tau^2}{3} + \frac{\theta_A \theta_C \tau^2}{3} + \frac{\theta_B \theta_C \tau^2}{3} - \frac{\theta_A \theta_B \theta_C \tau^3}{2}$$

8.19.2 Reversible state

In this mode all failures are revealed and equipment is subject to mean repair times. Let

$$P = \mu = \theta\tau_r$$

where θ is the revealed failure rate and τ_r is the mean repair time.

(a) *Probability/FDT inputs → probability/FDT output*

$$\text{Inputs} = P_A, P_B, P_C = \mu_A, \mu_B, \mu_C$$

$$\text{Output} = \mu_0$$

$$= P_A P_B + P_A P_C + P_B P_C - 2 P_A P_B P_C$$

or

$$\mu_A \mu_B + \mu_A \mu_C + \mu_B \mu_C - 2\mu_A \mu_B \mu_C$$

$$= \theta_A \theta_B (\tau_r)^2 + \theta_A \theta_C (\tau_r)^2 + \theta_B \theta_C (\tau_r)^2 - 2\theta_A \theta_B \theta_C (\tau_r)^3$$

$$= (\theta_A \theta_B + \theta_A \theta_C + \theta_B \theta_C)(\tau_r)^2 - 2\theta_A \theta_B \theta_C (\tau_r)^3$$

8.20 HIGH AND LOW DEMAND RATES IN RISK ASSESSMENT

It has been previously stated that providing process demand rates are low,

Hazard rate = (Process demand rate) × (FDT of protection system)

For practically all process systems the above relationship is seen to be valid because in safety terms process plants are necessarily designed and operated at low demand rate levels. A process plant which is subject to high dangerous demand rates or frequent spurious safe failure shutdowns is inherently unsafe and also uneconomic due to loss of production time caused by repairs and start-up procedures. It is also well known that process plants are generally more vulnerable to hazards during start-up periods.

It is pertinent when considering modelling principles to draw the attention of the reader to the effects and implications of high demand rates on the process hazard rate. Immediate clarity on the effect of high demand rates can be readily appreciated by reference to a hypothetical process which is subject to a level of demand rate which can be regarded as an approach to a continuous demand state. Assuming that the protective system responds correctly each time to a demand, it follows that the process hazard rate would approach a limiting value given by the failure rate of the protection system.

Proof testing, being a periodic outline routine applied to the protective system, seeks to verify that the system is free from dangerous unrevealed failures. Freedom from such failures can only be stated with reasonable certainty at the beginning or end of the test interval. When a process demand occurs during the test interval the defined hazard will ensue if the protective system should be in a fail danger state. Demands in safety engineering are assumed to be randomly distributed with time and it follows therefore that, on average, a demand will occur at the mid-point of the safety system proof test interval.

If demand rates are large such that the number of demands per proof test interval approaches or is in excess of unity, it follows that, on average, there is always a high probability of a demand being present between successive tests. In such circumstances, if the safety system responded according to its designed intent, it follows that a hazard state would be increasingly dependent on the **failure rate** of the protective system. Conversely, if the number of demands per test interval is low, then the assessed hazard rate is more conveniently dependent on the **failure probability** of the protective system.

8.20.1 High demand rates

When demand rates are high the hazard rate is given by protective system failure rate θ and the probability P_D of a demand occurring at the mid-point of the safety system proof test interval τ.

For the exponential or random distribution of demands:

$$P_D = 1 - e^{-0.5\tau D}$$

where D is the process demand rate. Hence

Hazard rate = (Failure rate of protection) × (Probability of demand

occurring at the mid-point of the proof test interval)

i.e.

$$H = \theta(1 - e^{-0.5\tau D})$$

$$= \theta P_D$$

8.20.2 Low demand rates

When demand rates are low, the hazard rate is more conveniently assessed on the failure probability on demand FDT and the process demand rate D:

Hazard rate = (FDT of protection) × (Process demand rate)

i.e.

$$H = \mu D$$

Table 8.2 Hazard rate comparisons

Demands D	Demands/τ	P_D	FDT μ	$H = \mu D$	$H = \theta P_D$
1.0	0.25	0.118	0.06	0.060	0.059
2.0	0.5	0.221	0.06	0.120	0.111
4.0	1.0	0.392	0.06	0.240	0.196
6.0	1.5	0.528	0.06	0.360	0.264
8.0	2.0	0.632	0.06	0.480	0.316
10.0	2.5	0.713	0.06	0.600	0.357
15.0	3.75	0.847	0.06	0.900	0.424
20.0	5.0	0.918	0.06	1.200	0.460
40.0	10.0	0.993	0.06	2.400	0.496

8.20.3 Comparisons of hazard evaluation methods

Table 8.2 lists hazard rates for a given typical system over a range of demands definable from low rates to those of high rates. The system safety parameters are defined as follows:

Failure rate of protection $= 0.5$ per annum

Proof test interval $\tau = 0.25$ years

FDT μ of protection $= 0.06$

The demand rate is variable over the range of 1–40 demands per annum.

The graph in Figure 8.7 shows that the hazard rates based on the FDT of the protective system and demand rate become increasingly invalid when the annual demand rate exceeds three per annum when referred to a protective system proof test interval of 0.25 years. The curves also show that over three demands per annum the hazard rate based on failure rate of the protective system and demand probability at the mid-point of the proof test interval increases exponentially to a limiting value equal to the failure rate of the protective system. Hence for high demands it is shown that hazard rate is given by demand probability over a period $\tau/2$ and the failure rate of the protective system.

A good illustration of the relative approaches to hazard rate quantification can be met by consideration of a vehicle brake system, whereby it is desired to assess the annual hazard rate for a particular vehicle in terms of loss of braking system through random dangerous failures. Let

Failure rate of the brake system $= 0.05$ faults per annum

Proof test interval based on legal requirements $= 1$ year

Demands (brake operations per annum) $= 10^4$

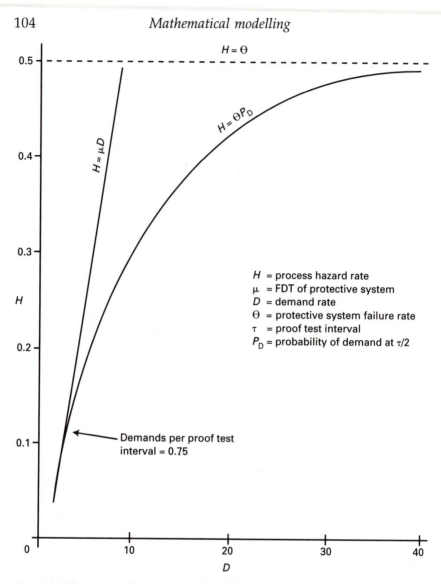

Figure 8.7 Hazard rate comparisons for high and low demand rates.

- Assess on basis of $H = \mu D$:

$$\mu = \frac{0.05 \times 1.0}{2} = 0.025$$

therefore

$$H = 0.025 \times 10^4 = 250 \text{ per annum}$$

which is clearly an invalid conclusion.

- Assess on basis of $H = \theta P_D$:

$$P_D = (1 - e^{-5000}) = 1.0$$

Noting that

$$-5000 = -\frac{D\tau}{2} \quad \text{(where } \tau = 1 \text{ year)}$$

therefore

$$H = 0.05 \times 1.0 = 0.05 \text{ per annum}$$

which is the failure rate of the brake system. Thus the loss of braking system is assessed as once in 20 years which represents a credible conclusion.

8.20.4 Concluding discussion

It may be concluded that the relationship $H = \theta P_D$ is universally valid for purposes of assessing process hazard rates. The method in practical terms is dependent on maintaining in service the performance rating of a given protective system in terms of its desired annual failure rate. Any projected enhancement of system failure rate is not easily achievable since all the components of the system, in order to meet changing criteria, are dependent on inherent subsystem failure rates set by type and manufacturer.

A hazard assessment approach through the relationship $H = \mu D$ is the normal accepted procedure used in safety engineering practice when demand rates are low, i.e. when demands/proof test interval are $\ll 1.0$.

At low demand rates both the stated hazard assessment procedures are therefore fully acceptable in terms of magnitude of risk. Since processes are designed and operated to meet low demand rate criteria, risk is almost always based on the relationship $H = \mu D$.

During service, reliability or conversely failure probability on demand is closely regulated by the frequency of proof testing such that increasing frequency will promote higher reliability performance rating. This is most practicably achieved by carrying out renewals where necessary at each proof test with the object of restoring the system to its initial fault-free state and hence its inherent failure rate requirement.

8.20.5 Summary

For low demand rates the preferred method of hazard rate evaluation is by taking the product of safety system FDT and process demand rate.

This method becomes increasingly less valid as demands per proof test interval approach unity.

In mathematical terms, evaluation of hazard rate by taking the product of safety system failure rate and demand probability at the midpoint of the proof test interval is universally valid.

It is more convenient in practical terms to maintain through periodic proof testing a protective system criterion in terms of mean fractional dead time rather than failure rate.

Subsystems with diverse failure rates can be operated and maintained at equitable failure probabilities through diverse periodic proof testing intervals.

9

Mathematical modelling of human failures

9.1 INTRODUCTION

Modelling of human reliability embraces the two facets of protective function and risk. The former concerns human operations which promote safety, whilst the latter deals with maloperations which may lead to hazard states.

In both modes, data on human behaviour relating to defined tasks are a necessary requirement when carrying out quantitative modelling. Task-related human data are obtainable from methods which are significant in number and variability and are normally generated by specialists in the field of human behaviour. The discussions which follow will therefore be confined to interpretation of human task-related data and how it is applied in modelling risk.

The main tenet of safety engineering philosophy is that the protective function will be segregated from that of the controlling function. This principle arises from a need to promote independence between the two and thereby to eliminate as far as possible commonalities which would initiate system demands and, at the same time, negate safeguard actions. Although this requirement is mostly satisfied in systems based on hardware and software, the human element is unique in that segregation of safety and controlling functions is rarely achieved. A process operator could accidentally initiate a dangerous demand and fail to take the necessary corrective action because of some underlying behavioural state on the part of the individual at the time. Hence it is emphasized that in modelling human behaviour the human element often bridges the boundary between safeguard actions and initiation of dangerous demands.

9.2 THE HUMAN ELEMENT

9.2.1 Safeguard actions

An alarm system which relies on an operator to take corrective action is significantly dependent on human reliability which is uncertain over a large range due to variable human reactions and transient extraneous influences. Such a system would clearly not be viable in major hazard situations.

When applying the human element to safety, the philosophy should be to improve the system protection by the reliability of the operator, however uncertain one may be of the data being used. For example if an automatic protective system incorporates an audible and visual alarm, then the shutdown action will be brought to the attention of the plant operator at the onset of the condition. The failure probability of the hardware can be predicted with a reasonable degree of confidence which would include the shutdown subsystem itself, e.g. automatic process isolating valves etc. The reliability of the shutdown subsystem is recognized as the least reliable part of any protective system when compared with input and logic interface elements. The alarm informs the operator that an automatic process shutdown is initiated and therefore he or she is expected to see the change in process state. If this does not occur due to failure of the shutdown element then the operator is able to take the necessary manual override action. Hence the reliability of the shutdown subsystem is enhanced by that of the human operator, however unreliable the person may be. In this scenario the human element contributes to the overall performance of the safeguard function.

9.2.2 Hazardous demands

Dangerous failures in process control systems constitute demands on the plant protective system. A control system which relies on a human operator to manually adjust a regulating element is very dependent on that person's reliability performance, which is subject to a wide range of influences. The system would not be acceptable in severe hazard situations where low frequency demands and hence risk is mandatory.

In order to eliminate uncertainty in operator reliability as well as meeting process response times, control systems incorporate automatic modulating elements such as control valves, heaters etc. Dangerous control system failures representing process demands are observable by the operator who would be able to take override manual corrective action. In this scenario, human element reliability performance, however uncertain, contributes to the achievement of lower demand rates and hence lower process system risk.

9.2.3 General comments

Reliance on human operator performance is significantly reduced by the employment of hardware and software systems which monitor and apply corrective actions and hence increase confidence levels in system reliability performance. However, it must be pointed out that automated systems of increasing complexity require maintenance and modification at higher technical levels of human influence. Elimination of dangerous human operator failures is therefore seen to be partly offset by failures of the human element in design and maintenance.

9.3 INTERPRETATION OF DATA

Analysts are often confronted with the task of assigning human data to either probability or failure rate. In the safeguard mode the reliability of the operator to carry out a given task, i.e. to respond, say, to an alarm condition, would clearly be in terms of probability of failure. In the demand mode the reliability of the operator would be viewed in terms of rate, i.e. annual frequency at which he or she would initiate the given demand. The application of a given piece of data poses the question of how to use it as a probability or rate. This can be resolved by considering the dimensional criteria terms under which probability and rate elements must comply.

Probability is a dimensionless number and therefore has no element of time. It may be described as the chance of failure to carry out a given task in a required manner, for example one chance in 1000, i.e. 10^{-3}.

Rate, however, has a dimension of inverse time denoted by T^{-1} which typically could be described in terms of events or demands per annum. It follows that if, for example, the available data state that the specific task to be carried out by a type of operator will fail probably once in 1000 times then such data will need to be modified by inverse time T^{-1} in order to derive the rate form.

Suppose an operator carries out a given task 200 times per annum such that any one if incorrect would constitute a demand and that his or her probability of failure to carry out the task correctly is 1 in 1000.

$$\text{Operations/annum in dimensional terms} = \frac{\text{Tasks}}{\text{Time}} = \frac{[1]}{[T]} = T^{-1}$$

noting that task is a dimensionless quantity.

Annual demand rate = (Operations per annum)

$$\times \text{ (Failure probability per operation)}$$

$$= \frac{200}{1 \text{ year}} \times \frac{1}{1000} = 2 \times 10^{-1} \text{ per annum}$$

In dimensional terms,

$$\frac{\text{Tasks}}{\text{Time}} \times \text{Probability} = \frac{[1]}{[T]} \times [1] = T^{-1}$$

Therefore 2×10^{-1} per annum is confirmed as a rate term quantity.
    In general terms, let

f = annual frequency of a given task to be carried out by the human
    operator

P_H = failure probability of operator when carrying out the task

θ_H = annual human failure rate

Then

$$\theta_H = f P_H$$

9.4 ASPECTS OF HUMAN FAILURE

Human beings are by nature not subject to proof testing or mean repair times and therefore any given human failure, whether in terms of rate or probability, is directly applicable to both the partly reversible and the reversible state modes. Human failure probability, although known to be transiently variable, is normally regarded as being constant over the duty period and is therefore by definition a mean fractional dead time.

9.5 HUMAN AND HUMAN–EQUIPMENT FAILURE COMBINATIONS AT LOGIC GATES

In order to demonstrate the principles of combinations at OR gates, AND gates and majority voting gates, inputs will be described in terms of input sets which consist of three elements. The methods to be demonstrated will enable input sets of smaller or greater numbers of member elements to be transposed to single probability or rate output expressions. The universal Boolean deterministic relationships in respect of three elements are stated respectively as follows for the three logic gate types.

- Inputs to OR gate $= A, B, C$

$$\text{Output} = A \text{ OR } B \text{ OR } C = A + B + C$$

- Inputs to AND gate $= A, B, C$

$$\text{Output} = A \text{ AND } B \text{ AND } C = ABC$$

- Inputs to majority voting gate $= A, B, C$

$$\text{Output} = AB \text{ OR } AC \text{ OR } BC = AB + AC + BC$$

9.5.1 Symbols

$\theta_H =$ human failure rate

$\theta_D =$ equipment failure rate in the unrevealed mode

$\theta_R =$ equipment failure rate in the revealed mode

$\tau =$ proof test interval relating to the unrevealed failure mode

$\tau_r =$ mean repair time relating to the revealed failure mode

$P_H =$ human failure probability $= \mu_H$

$\mu_H =$ human failure probability on demand $= P_H$

$\mu_D =$ equipment mean fractional dead time in the unrevealed failure mode

$P_R =$ equipment failure probability in the revealed failure mode $= \mu_R$

$\mu_R =$ equipment mean fractional dead time in the revealed failure mode $= P_R$

$P_0 =$ gate output probability

$\mu_0 =$ gate output mean fractional dead time

$\theta_0 =$ gate output rate

$P_D =$ equipment failure probability in the unrevealed failure mode

$t =$ time

Also

$$P_D = \theta_D \tau$$

$$\mu_D = \frac{1}{\tau} \int_0^\tau P_D \, dt$$

and

$$P_R = \mu_R = \theta_R \tau_r$$

9.5.2 Failure state modes

All input sets will be referred to both the partly reversible and the reversible state failure modes in which equipment hardware, where pertinent, is an operational system requirement in combination with those of human input elements. All derived expressions will be given in probabilistic form.

9.6 COMBINATIONS AT THE LOGICAL OR GATE

These will be considered in terms of either rates or probabilities, not combinations of both, in order to comply with dimensional criteria.

9.6.1 Rates

(a) Partly reversible state

- Human input set:

$$\text{Input} = \theta_{H1}, \theta_{H2}, \theta_{H3}$$

$$\text{Output } \theta_0 = \theta_{H1} + \theta_{H2} + \theta_{H3}$$

  i.e. the arithmetic sum of the rates.
- Human and equipment input set:

$$\text{Input} = \theta_{H1}, \theta_{D1}, \theta_{D2}$$

  i.e. one human and two equipment elements.

$$\text{Output } \theta_0 = \theta_{H1} + \theta_{D1} + \theta_{D2}$$

  i.e. the arithmetic sum of the rates.

(b) Reversible state

- Human input set:

$$\text{Input} = \theta_{H1}, \theta_{H2}, \theta_{H3}$$

$$\text{Output } \theta_0 = \theta_{H1} + \theta_{H2} + \theta_{H3}$$

  i.e. identical to that in section 9.6.1(a).

- Human and equipment input set:

$$\text{Input} = \theta_{H1}, \theta_{R1}, \theta_{R2}$$

i.e. one human and two equipment elements.

$$\text{Output } \theta_0 = \theta_{H1} + \theta_{R1} + \theta_{R2}$$

i.e. the arithmetic sum of the rates.

9.6.2 Probabilities

(a) Partly reversible state

Probability inputs → probability output

- Human input set:

$$\text{Input} = P_{H1}, P_{H2}, P_{H3}$$

Output is

$$P_0 = \mu_0 = P_{H1} + P_{H2} + P_{H3} \quad \text{(neglecting higher-order terms)}$$
$$P_0 = \mu_0 = P_{H1} + P_{H2} + P_{H3} - (P_{H1}P_{H2} + P_{H1}P_{H3} + P_{H2}P_{H3})$$

(neglecting third-order term)

- Human and equipment input set:

$$\text{Input} = P_H, P_{D1}, P_{D2}$$

Output is

$$P_0 = P_H + P_{D1} + P_{D2} \quad \text{(neglecting higher-order terms)}$$
$$P_0 = P_H + P_{D1} + P_{D2} - (P_H P_{D1} + P_H P_{D2} + P_{D1}P_{D2})$$

(neglecting third-order term)

Probability inputs → FDT output

- Human input set:

$$\text{Input} = P_{H1}, P_{H2}, P_{H3}$$

Output is

$$P_0 = \mu_0 = P_{H1} + P_{H2} + P_{H3} \quad \text{(neglecting higher-order terms)}$$
$$P_0 = \mu_0 = P_{H1} + P_{H2} + P_{H3} - (P_{H1}P_{H2} + P_{H1}P_{H3} + P_{H2}P_{H3})$$

(neglecting third-order term)

- Human and equipment input set:

$$\text{Input} = P_H, P_{D1}, P_{D2}$$

Output is

$$\mu_0 = P_H + \mu_{D1} + \mu_{D2} \quad \text{(neglecting higher-order terms)}$$
$$\mu_0 = P_H + \mu_{D1} + \mu_{D2} - [P_H\mu_{D1} + P_H\mu_{D2} + \text{FDT}(\mu_{D1}\mu_{D2})]$$
$$= P_H + \mu_{D1} + \mu_{D2} - \left(P_H\mu_{D1} + P_H\mu_{D2} + \frac{4\mu_{D1}\mu_{D2}}{3}\right)$$

(neglecting third-order term)

(b) Reversible state

Probability inputs → probability output
Note that the equations below also describe FDT inputs → FDT output.

- Human input set:

$$\text{Input} = P_{H1}, P_{H2}, P_{H3}$$

Output is

$$P_0 = P_{H1} + P_{H2} + P_{H3} \quad \text{(neglecting higher-order terms)}$$
$$P_0 = P_{H1} + P_{H2} + P_{H3} - (P_{H1}P_{H2} + P_{H1}P_{H3} + P_{H2}P_{H3})$$

(neglecting third-order term)

- Human and equipment input set:

$$\text{Input} = P_H, P_{R1}, P_{R2}$$

Output is

$$P_0 = P_H + P_{R1} + P_{R2} \quad \text{(neglecting higher-order terms)}$$
$$P_0 = P_H + P_{R1} + P_{R2} - (P_HP_{R1} + P_HP_{R2} + P_{R1}P_{R2})$$

(neglecting third-order term)

9.7 COMBINATIONS AT THE LOGICAL AND GATE

Rates or probabilities or combinations of both may be taken through AND gates without violating the requirements of dimensional criteria. However, it should be noted that rates when combined through AND gates need to be modified by probability factors in order to maintain dimensional validity of output rate.

9.7.1 Rates

The principles of human failure rate transitions through AND gates are identical to those already discussed in Chapter 8 for equipment-based input sets. The human failure mode is not subject to either proof testing or repair regimes and is therefore directly applicable without modification to both the partly reversible and reversible failure modes.

(a) Partly reversible state

- Human input set:

$$\text{Input} = \theta_{H1}, \theta_{H2}, \theta_{H3}$$

  Output is given by

  (H1 fails) $\times$ (Probability that H2 and H3 have failed) +

  (H2 fails) $\times$ (Probability that H1 and H3 have failed) +

  (H3 fails) $\times$ (Probability that H1 and H2 have failed)

  Hence

$$\theta_0 = \theta_{H1}P_{H2}P_{H3} + \theta_{H2}P_{H1}P_{H3} + \theta_{H3}P_{H1}P_{H2}$$

- Human and equipment input set:

$$\text{Input} = \theta_H, \theta_{D1}, \theta_{D2}$$

  Output is

$$\theta_0 = \theta_H \times \text{FDT}(P_{D1}, P_{D2}) + \theta_{D1}\mu_{D2}P_H + \theta_{D2}\mu_{D1}P_H$$

$$= \theta_H \frac{1}{\tau} \int_0^\tau P_{D1}P_{D2} \, dt + \theta_{D1}\mu_{D2}P_H + \theta_{D2}\mu_{D1}P_H$$

$$= \frac{\theta_H\theta_{D1}\theta_{D2}\tau^2}{3} + \theta_{D1}\mu_{D2}P_H + \theta_{D2}\mu_{D1}P_H$$

(b) Reversible state

- Human input set:

$$\text{Input} = \theta_{H1}, \theta_{H2}, \theta_{H3}$$

$$\text{Output } \theta_0 = \theta_{H1}P_{H2}P_{H3} + \theta_{H2}P_{H1}P_{H3} + \theta_{H3}P_{H1}P_{H2}$$

- Human and equipment input set:

$$\text{Input} = \theta_H, \ \theta_{R1}, \ \theta_{R2}$$

$$\text{Output } \theta_0 = \theta_H P_{R1} P_{R2} + \theta_{R1} P_H P_{R2} + \theta_{R2} P_H P_{R1} \quad (\text{note that } P_R = \mu_R)$$

9.7.2 Probabilities

(a) Partly reversible state

Probability inputs → probability output

- Human input set:

$$\text{Input} = P_{H1}, \ P_{H2}, \ P_{H3}$$

$$\text{Output } P_0 = P_{H1} P_{H2} P_{H3}$$

- Human and equipment input set:

$$\text{Input} = P_H, \ P_{D1}, \ P_{D2}$$

$$\text{Output } P_0 = P_H P_{D1} P_{D2}$$

Probability inputs → FDT output

- Human input set:

$$\text{Input} = P_{H1}, \ P_{H2}, \ P_{H3}$$

$$\text{Output } \mu_0 = P_{H1} P_{H2} P_{H3}$$

- Human and equipment input set:

$$\text{Input} = P_H, \ P_{D1}, \ P_{D2}$$

$$\text{Output } \mu_0 = P_H \times \text{FDT}(P_{D1} P_{D2})$$

$$= P_H \times \frac{1}{\tau} \int_0^\tau P_{D1} P_{D2} \ dt$$

$$= P_H \times \frac{4\mu_{D1}\mu_{D2}}{3}$$

(b) Reversible state

Probability/FDT inputs → probability/FDT output

- Human input set:

$$\text{Input} = P_{H1}, P_{H2}, P_{H3}$$

$$\text{Output } P_0 = \mu_0 = P_{H1}P_{H2}P_{H3}$$

- Human and equipment input set:

$$\text{Input} = P_H, P_{R1}, P_{R2}$$

$$\text{Output } P_0 = \mu_0 = P_H P_{R1} P_{R2}$$

9.7.3 Rates and probabilities

(a) Partly reversible state

- One human rate and two equipment probabilities:

$$\text{Input} = \theta_H, P_{D1}, P_{D2}$$

$$\text{Output } \theta_0 = \theta_H \times \text{FDT}(P_{D1}P_{D2})$$

$$= \theta_H \times \frac{1}{\tau} \int_0^\tau P_{D1}P_{D2} \, dt$$

$$= \theta_H \times \frac{4\mu_{D1}\mu_{D2}}{3}$$

- Two human rates and one equipment probability:

$$\text{Input} = \theta_{H1}, \theta_{H2}, P_D$$

$$\text{Output } \theta_0 = (\theta_{H1}P_{H2} + \theta_{H2}P_{H1}) \frac{1}{\tau} \int_0^\tau P_D \, dt$$

$$= (\theta_{H1}P_{H2} + \theta_{H2}P_{H1})\mu_D$$

- One human probability and two equipment rates:

$$\text{Input} = P_H, \theta_{D1}, \theta_{D2}$$

$$\text{Output } \theta_0 = (\theta_{D1}\mu_{D2} + \theta_{D2}\mu_{D1})P_H$$

(b) Reversible state

- One human rate and two equipment probabilities:

$$\text{Input} = \theta_H, P_{R1}, P_{R2}$$

$$\text{Output } \theta_0 = \theta_H P_{R1} P_{R2}$$

- Two human rates and one equipment probability:

$$\text{Input} = \theta_{H1}, \theta_{H2}, P_R$$

$$\text{Output } \theta_0 = (\theta_{H1}P_{H2} + \theta_{H2}P_{H1})P_R$$

- One human probability and two equipment rates:

$$\text{Input} = P_H, \theta_{R1}, \theta_{R2}$$

$$\text{Output } \theta_0 = P_H(\theta_{R1}P_{R2} + \theta_{R2}P_{R1})$$

9.8 COMBINATIONS IN MAJORITY VOTING LOGIC

Human operations orientated to majority voting systems may encompass protective function and risk modes. In the former, voting is a rarely used practice in the field of human operations at the process plant level and is therefore to be most commonly found in equipment dedicated to safety. A prime reason for the choice of equipment rather than the human element is that safety-orientated reaction times necessary to meet most process demands are more reliably met in equipment systems than reliance on human majority decisions.

Human beings are subject to errors of judgement in all walks of life, including process plant operations, where consequences may vary from the insignificant to that of the major hazard. On many processes, manual operational procedures may, if carried out erroneously, cause the process system to proceed towards a dangerous state which is recognized as a demand. This constitutes a risk mode and in such scenarios it is not uncommon to employ checking regimes carried out by more than one human element, i.e. to base the action on a majority judgement.

Majority voting systems consisting of mixed human and equipment elements are very rarely found in practice and therefore in the considerations which follow only the human element will be considered. Also for method demonstration a 'two-out-of-three' human-based failure majority voting system will be the subject of illustration.

Referring to section 8.6 the Boolean expression for a two-out-of-three failure voting gate is given as

$$AB + AC + BC$$

This expression, being universal to both rates and probabilities, will now be used to derive the respective equivalent probabilistic expressions which follow, where θ_A represents the failure rate of human operator A etc. and P_A the failure probability of human operator A etc.

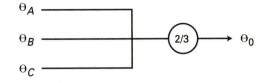

Figure 9.1 Two-out-of-three human voting in terms of system rate.

9.8.1 Rates

Figure 9.1 refers to two-out-of-three human voting in terms of system rate. Referring to the Boolean equation $AB + AC + BC$, the gate output is

$$\theta_0 = (\theta_A P_B + \theta_B P_A) + (\theta_A P_C + \theta_C P_A) + (\theta_B P_C + \theta_C P_B)$$

9.8.2 Probabilities

Figure 9.2 refers to two-out-of-three human voting in terms of system probability. Referring to the Boolean equation $AB + AC + BC$, the gate output is

$$P_0 = P_A P_B + P_A P_C + P_B P_C$$

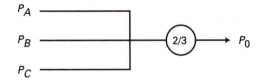

Figure 9.2 Two-out-of-three human voting in terms of system probability.

10

Modelling common-mode failures

10.1 INTRODUCTION

This type of failure most commonly called 'common mode' is also referred to as 'common cause' or 'dependent failure' and probably represents the most significant limiting factor in the achievement of low-risk criteria in safety engineering. Common-mode failure embraces both equipment hardware and management software with the thread of human error running throughout.

10.2 DEFINITION OF TERMS

10.2.1 Common-mode failure

This is defined as a type of redundant system failure which is attributable to a common or single cause.

10.2.2 Dependent failure

This is defined as a single failure event which is capable of transmission to each element or channel of a given redundant system.

10.2.3 Common-cause failure

This is defined as a specific condition which may result in a single failure event and which would be capable of causing each element or channel of a redundant system to fail.

10.2.4 Critical period

Multiple failures through common cause of elements or channels of a redundant system will result in overall system failure if they occur within a time interval known as the critical period.

10.2.5 Concurrency

When multiple failures through common cause of elements or channels of a redundant system occur within the critical period, the individual element or channel failures are said to be concurrent. System failure in concurrency mode is normally incapable of prevention by human intervention or scheduled changes in process operation.

10.2.6 Non-concurrency

When multiple failures through common cause of elements or channels of a redundant system occur over a given time interval greater than the critical period, the individual element or channel failures are said to be non-concurrent. System failure in non-concurrency mode is capable of prevention by human intervention or scheduled changes in process operation.

10.3 SYMBOLS

θ_{CM} = common-mode failure rate component of a total failure rate

θ_I = independent failure rate component of a total failure rate

P_{CM} = common-mode failure probability component of a total failure probability

P_I = independent failure probability component of a total failure probability

P_T = total failure probability

P_0 = system failure probability of a redundant system

θ = failure rate

τ = proof test interval

μ = mean fractional dead time

β = ratio factor common mode/total failure mode

10.4 THE NATURE OF COMMON-MODE FAILURE IN ASSESSMENT

The modelling of system failures due to common causes uses the same fundamentals as those for independent modelling. Common-mode failures may therefore be expressed in terms of failure rate, probability or mean fractional dead time and comply fully with the classical approach which has long been established for the independent model mode.

In order to extend the common-mode aspect into the classical assessment methods it is necessary to acquire data and to segregate and recognize it in relation to that of the independent whilst at the same time be able to integrate the two into an overall system failure model.

10.4.1 Common-mode failure data

Data specific to common-mode failure are very difficult and costly to identify and hence to acquire, so are not readily available from data bank sources. The concurrency factor in field data may be regarded as the principal problem in data analysis. Repairs or renewals over longer time intervals make it most difficult to relate failures to a common cause, particularly if there is a turnover of maintenance personnel and if maintenance records are not specifically organized to recognize common factors in successive failures.

General field data supplied by data banks for any single subsystem or equipment should always be considered, unless otherwise specified, as raw data in that they comprise independent and dependent failure contributors which make up the total failure rate. The overriding problem in risk assessment is to differentiate between the dependent failure and that of the independent failure. To this end there have been many modelling methods, mostly theoretically based, which lack a practicable engineering approach to a satisfactory understanding and solution.

The modelling method to be presented will put forward a methodology which is based on practical experience using limited common-mode empirical data which are most conveniently processed through the medium of fault trees.

10.4.2 Defence against common-mode failure

Defence against common-mode failures is dependent entirely on diversity. In the hypothetical situations of two diversity extremes of 0% and 100%, common-mode failures would be inherently at a maximum in the former and entirely eliminated at the latter high level. In practice 100% diversity is not achievable since commonalities always exist in the forms of environment, maintenance, process management, manufacture etc., extending also to plant site influences such as fire, explosion or changes in working practice.

Limited empirical common-mode failure data given in UKAEA publication SRD R147, *A Study of Common Mode Failures*, suggests that the ratio of common-mode probability or rate to overall probability or rate extends over the range of 0.1 to 0.001 corresponding to a diversity range in service of 0–100%.

10.4.3 Common-mode influence on independent failure modelling

In practice, independent failure modelling is always assumed to use independent failure data. All data, and in particular field data in respect of single items, are collected and expressed in a total form constituted principally by safe and dangerous independent failure mode elements accompanied by additional elements related to dependent-mode failures. Common-mode failure elements particularly in respect of dangerous failures can be due to a number of causes such as manufacturing, errors in application, operational environments and human-based faults in testing, maintenance, commissioning and recommissioning.

In non-redundant systems, common-mode modelling considerations are not applied, but nevertheless the common-mode element remains present and manifests itself as a component of the independent failure mode. It is only when redundancy is employed that the common-mode element, taken in isolation, contributes to the overall mathematical model. This therefore calls into question the validity of independent failure modelling.

In order to examine the influence of the common-mode component in independent modelling it is convenient to consider two-channel and three-channel safety logics presented in failure mode. The considerations will be based on a mean ratio factor of common-mode failure probability to total single-channel failure probability over the nonlinear ratio range of 0.1 to 0.001 and hence to yield a mean diversity across the given logics.

Section 10.7 states that

$$\beta = e^{-(0.046D_p+2.3)}$$

where β is the ratio of common mode to total failure mode and D_p is the percentage diversity across redundant channels.

Mean value β_{av} over the range of 0–100% diversity is given by

$$\beta_{av} = \frac{1}{100} \int_0^{100} e^{-(0.046D_p+2.3)} \, dD_p$$

From whence

$$\beta_{av} = 0.022$$

$$0.022 = e^{-(0.046D_p + 2.3)}$$

$$\log_e 0.022 = -0.046D_p - 2.3$$

from which $D_p = 33\%$.

Let:

P_T = total dangerous failure probability of single channel

P_I = independent dangerous failure probability of single channel

P_{CM} = dependent failure contributor from single channel

Also $P_{CM} = \beta P_T$, hence

$$P_I = P_T - P_{CM}$$

Errors in independent failure modelling due to common mode will now be considered, commencing with the most simple redundancy logic of 2/2.

(a) Two-channel system

Figure 10.1 shows the independent model which uses the total failure probability values P_T consisting of independent and dependent components, i.e. $P_T = P_I + P_{CM}$. The dependent failure elements are ignored and hence the classical approach to independent failure modelling is safe and pessimistic. The true independent model would, however, be as shown in Figure 10.2. Comparing the two equations in Figures 10.1

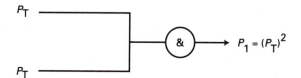

Figure 10.1 Independent model for two-channel system using total failure probability values.

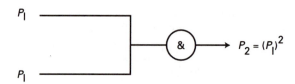

Figure 10.2 True independent model.

and 10.2 it is seen that Figure 10.1 is in error, which is safe and pessimistic in safety terms. The percentage error on the system of Figure 10.1 over that of the true value given by Figure 10.2 is

$$E = \frac{(P_T)^2 - (P_I)^2}{(P_I)^2} \times 100\%$$

$$E = \left(\frac{(P_T)^2}{(P_I)^2} - 1\right) \times 100\% \qquad (10.1)$$

(b) Three-channel system

Figure 10.3 shows the classical independent failure probability model for a three-channel redundant system logic based on total failure probability P_T. The model includes contributions from common-mode failure contained in each of the channels and is therefore safe and pessimistic. The true independent model is given in Figure 10.4. Comparing the two equations in Figures 10.3 and 10.4 it is seen that Figure 10.3 is in error, which is safe and pessimistic in safety terms. The percentage error on the system of Figure 10.3 over that of the true value given by Figure 10.4 is

$$E = \frac{(P_T)^3 - (P_I)^3}{(P_I)^3} \times 100\%$$

$$E = \left(\frac{(P_T)^3}{(P_I)^3} - 1\right) \times 100\% \qquad (10.2)$$

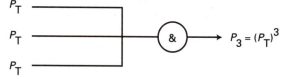

Figure 10.3 Independent model for three-channel system using total failure probability values.

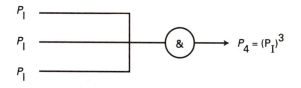

Figure 10.4 True independent model.

(c) Errors for two-, three- and four-channel systems

Table 10.1 gives errors generated by two-, three- and four-channel redundant systems over those of the true models. The error values are seen to be reliant only on the degree of redundant logic and can be shown to be independent of the magnitude of single-channel total failure probability.

Table 10.1 Percentage diversity/redundant logic error margins

Percentage diversity	Percentage error in failure logic		
	2/2	3/3	4/4
0	+23.4	+37	+52
20	+8.5	+13	+17
33	**+4.5**	**+6.9**	**+9.3**
40	+3.3	+5.0	+6.7
60	+1.2	+3.1	+4.1
80	+0.6	+0.9	+1.2
100	+0.2	+0.3	+0.4

(d) Conclusions

Classical quantified independent mathematical modelling is based on empirical data which are always assumed to be totally independent in nature and therefore ignores the dependent failure component in each of the data items. This results in pessimistic safe errors varying from 0.2% to 52% in the case of redundant logics up to the fourth order, as shown in Table 10.1. Applying the range of errors to redundant systems having similar channel failure probabilities, it is seen that the errors are reasonably acceptable. For example, if individual channels having 0% mutual diversities in a given logic each have failure probabilities of, say 10^{-1}, then the true values for each of the redundant logics in the table would be 8.1×10^{-3}, 7.3×10^{-4} and 6.6×10^{-5}. The assessed values, using empirical data, would be 10^{-2}, 10^{-3} and 10^{-4} respectively. Hence the true values can be reasonably regarded as acceptable in view of their pessimistic natures. As diversities increase, the errors become significantly smaller from and beyond the 33% statistical average diversity level shown in bold in Table 10.1.

10.5 THE LOGICAL APPROACH TO COMMON-MODE FAILURE ASSESSMENT

Common-mode failure mechanisms are related only to Boolean 'AND' logic scenarios. Redundancy in systems, whether protective or plant

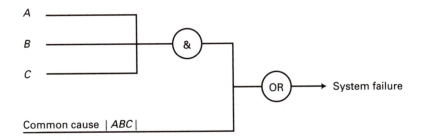

Figure 10.5 Common-mode failure logic for three independent failure inputs.

operational, is always employed in order to achieve higher reliabilities for safety or availability when in service. In the protective or plant system modes, *A* AND *B* AND *C* etc. must fail in order to bring about total system failure and therefore, from the definition of common-mode failure, the logic can be stated as in Figure 10.5, where *A*, *B* and *C* are Boolean inputs.

Hence when system failure is expressed in terms of Boolean ANDs, each group will generate a single common-mode system failure element which is only present if *A*, *B* and *C* are present and failures are concurrent.

Specifically in any risk or safety assessment the generation and hence analysis of dependent failures is enabled by each minimal cut set in the total listing derived from a fault tree which describes overall system failure.

10.6 COMMON-MODE ANALYSIS FROM INDEPENDENT MINIMAL CUT SETS

In order to demonstrate assessment of failures due to common causes in terms of probabilities or rates it is proposed to consider failure mode independent minimal cut sets from the single-order to fourth-order terms.

10.6.1 Single-order independent cut set

Let this consist of a Boolean element *A* in isolation such that

$$\text{System failure} = A$$

Since no other failure elements are present it follows that failure of *A* is due only to random failures which are behaviourally independent in character. The common-mode component of independent failure rate manifests itself as part of the random independence and only becomes active in a dependent failure sense in the presence of other redundant members, should it be included in a multi-element set.

10.6.2 Second-order independent cut set

This is defined as elements *A* AND *B*, i.e.

$$\text{System independent failure} = AB$$

Hence there can only be one dependent failure which will fail both *A* and *B*, noting that its occurrence must be at least equal to the most reliable of the two elements. To put this statement into clearer perspective, let *A* be the element with the most favourable failure rate, say θ_A. Therefore the dependent failure which would affect *A* and *B* must be at least equal to or better than the independent failure of *A*, i.e.

$$\text{Common-mode failure rate } \theta_{\text{CM}} < \theta_A$$

or

$$\text{Common-mode failure probability } P_{\text{CM}} < P_A$$

Let $|AB|$ denote dependent failure across independent elements *A* and *B*. Hence a second-order independent minimal cut set produces one single-order common-mode term $|AB|$. The logic is expressed in Figure 10.6.

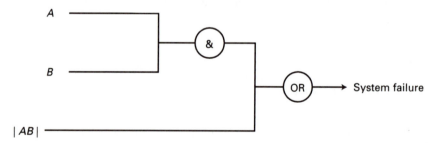

Figure 10.6 Failure logic for second-order independent cut set and derived common mode.

10.6.3 Third-order independent cut set

This is defined as *ABC* and generates dependent failures as follows:

- Dependent failure across *ABC*;
- (Dependent failure across *AB*) × (Independent failure of *C*);
- (Dependent failure across *AC*) × (Independent failure of *B*);
- (Dependent failure across *BC*) × (Independent failure of *A*).

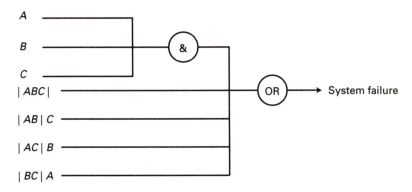

Figure 10.7 Failure logic for third-order independent cut set and derived common mode.

This gives the following Boolean common-mode expression for cut set *ABC*:

$$\text{Common-mode failure} = |ABC| + |AB|C + |AC|B + |BC|A$$

Hence a third-order independent minimal cut set produces four common-mode cut sets comprising one single-order and three second-order common-mode terms. The relevant logic is given in Figure 10.7.

Note that dependent failure $|ABC|$ is not the same dependent failure $|AB|$ OR $|AC|$ OR $|BC|$, although common-cause influences in each of the second-order common-mode sets will contain selective independent failure influences from the group population associated with the single-order set *ABC*.

10.6.4 Fourth-order independent cut set

This is defined as elements *ABCD*, hence the common-mode cut sets are

$$|ABCD| + |ABC|D + |ABD|C + |ACD|B + |BCD|A + |AB|CD$$
$$+ |AC|BD + |AD|BC + |BC|AD + |BD|AC + |CD|AB$$

i.e. a fourth-order independent minimal cut set generates 11 common-mode cut sets.

10.6.5 Approximations in common-mode cut sets

The above common-mode cut set derivations show that their numbers would be considerable in any risk or reliability assessment where larger populations of independent minimal cut sets are present. To evaluate the contribution from each common-mode cut set would not be justifiable since contributions become increasingly insignificant with higher orders. Hence it is convenient at this stage to define significance, and for this purpose a four-element independent minimal cut set *ABCD* will be considered in terms of its derived single-order and second-order common-mode cut sets.

(a) Single-order common-mode cut set

This is defined as |*ABCD*| which refers to a dependent fault which will result in common-mode failure of elements *ABCD*.

Probability mode

> Common-mode failure probability across elements *ABCD*
> > = [Ratio factor (0.1 to 0.001 subject to diversity)]
> > > ×(Failure probability of most available element from *ABCD*)

Hence the single-order common-mode set failure probability is seen to be the product of two factors, each of which are fractional.

Rate mode

> Common-mode failure rate across elements *ABCD*
> > = [Ratio factor (0.1 to 0.001)]
> > > ×(Failure rate of most available element from *ABCD*)

In the rate mode one factor will always be fractional, namely the ratio factor, whilst the rate term factor, although generally fractional, can sometimes equal or exceed unity.

(b) Second-order common-mode cut set

The independent minimal cut set *ABCD* provides a total of four common-mode second-order sets, namely |*ABC*|*D*, |*ABD*|*C*, |*ACD*|*B* and |*BCD*|*A*, any one of which, if present, will cause failure of the system *ABCD*.

Probability mode

Common-mode failure probability arising from, say |*ABC*|*D* is given by

> (Dependent failure probability across *ABC*)
> > × (Independent failure probability *D*)

$$= [\text{Ratio factor (0.1 to 0.001)}]$$

$\times$ (Failure probability of most available element from ABC)

$\times$ (Independent failure probability D)

Hence the second-order common-mode set failure probability is the product of three factors, each of which is fractional and may therefore generally be regarded as insignificant to that of the associated single-order common-mode set. In the event of uncertainty, a reasonable recourse would be to multiply the value of a single set by the number of second-order common-mode sets, namely four in the case of the minimal independent cut set $ABCD$ and the result added if necessary to that of the single-order common-mode set. Hence the common-mode approximation for the minimal independent cut set $ABCD$ would be

$$|ABCD| + 4(|ABC|D)$$

Rate mode

Let the independent minimal cut set $ABCD$ contain a single rate element defined as that of A, all other elements BCD being associated with probabilities. It can be seen from the probability mode method above that the element A can appear within the common-mode group $|ABC|D$, or outside it, $|BCD|A$.

Consider second-order set $|ABC|D$:

Common-mode failure rate

$$= \text{(Ratio factor)}$$

$\times$ (Most favourable failure rate from ABC)

$\times$ (Independent failure probability D)

Note that B and C, being probabilities, are derived from failure rates.

Approximation of second-order common-mode sets by the method recommended for the probability mode above is not generally recommended in the rate mode since rates are not always fractional.

Consider second-order set $|BCD|A$:

Common-mode failure rate

$$= \text{(Ratio factor)}$$

$\times$ (Most favourable independent failure probability from BCD)

$\times$ (Failure rate A)

In both of the above cases the common-mode failure rate is given by the product of two fractional quantities and a single rate term.

(c) Conclusion

Table 10.2 lists numbers of common-mode cut sets related to a range of independent minimal cut sets from second order to sixth order, which is representative of most assessment studies. In the majority of cases it is not necessary to take into account the common-mode contributors from second-order common-mode sets, but where reasonable doubt exists concerning significance, approximations may be carried out as previously described. Third-order and higher common-mode sets may be ignored.

Table 10.2 Independent minimal cut-set/common-mode cut-set derivations

Independent minimal cut-set orders	Common-mode set orders					Totals
	1st	*2nd*	*3rd*	*4th*	*5th*	
1st	–	–	–	–	–	0
2nd	1	–	–	–	–	1
3rd	1	3	–	–	–	4
4th	1	4	6	–	–	11
5th	1	5	10	6	–	22
6th	1	6	15	10	6	38

10.7 COMMON-MODE ASSESSMENT METHODOLOGY

Independent minimal cut sets generated from fault trees are seen to be the source of those common-mode failure contributors which make up the overall system common-mode failure rate or probability. All such cut sets are initially expressed in Boolean terms for subsequent transfer to the probabilistic form in terms of rates or probabilities. The minimal cut sets therefore require a methodology by which the common-mode contributor elements can be extracted and thence evaluated.

The basis of the method by which common-mode failure contributors are quantified is known as the β (beta) factor model which was developed by the General Atomic Company in the USA for the quantification of the reliability of redundant systems and is described in UKAEA publication SRD R146, *A Study of Common Mode Failures*.

The beta factor relates to the ratio (previously defined) of the common-mode failure rate or probability of the most available redundant element of a given group to its total failure rate or probability. The model, based on field data, suggests that a range of beta values from 0.001 to 0.1 could be applicable from almost fully diverse to non-diverse redundant systems.

Diversity exists not only in equipment type but also in manufacture, supply, function, design, maintenance, process operation and management software. In any rigorous risk or safety assessment each independent failure minimal cut set could be analysed for all degrees of mode diversities. However, reasonable assessment may be based principally on degrees of diversity based on subsystem type supplemented where necessary by subjective judgements on the presence of the other named factors.

Limited field data quoted in UKAEA SRD R146 also suggests that concurrency of dependent failures within a critical time interval of say 8–12 hours will account for 20–25% of the overall common-mode failure rate or probability. Since all observed field data contain common-mode elements, 'beta' is defined as

$$\beta = \frac{P_{CM}}{P_I + P_{CM}}$$

or in terms of failure rate,

$$\beta = \frac{\theta_{CM}}{\theta_I + \theta_{CM}}$$

In assessing P_{CM} (or θ_{CM}) the factor is applied to the most available redundant element in the system. The magnitude of the dependent failure contribution arising from a given common-mode cut set is based on the degree of diversity present across the dependent element group in the given cut set in accordance with the law

$$\beta = e^{-(0.046D_p + 2.3)}$$

where D_p is the percentage diversity across the given cut-set elements.

10.7.1 Evaluation of percentage diversity across a minimal cut set

(a) Degree of diversity

In any parent cut set where the elements are 100% mutually diverse then any pair of elements would be diverse to each other and hence would contribute a degree of diversity of 1 to the parent cut set. Hence 50% diversity would correspond to a degree of diversity of 0.5.

(b) Cut-set element pair

This is defined as any pair of elements taken from a parent set. The number of element pairs generated from a given set is given by

$$S = \frac{n_s(n_s - 1)}{2}$$

where S is the number of element pairs and n_s is the number of elements in a given set.

(c) Total element pairs in parent cut set

The total number of element pairs generated by a parent minimal cut set is defined as S_p.

(d) Total active element pairs in parent cut set

An active element pair is defined as any element pair, taken from a parent minimal cut set, which has a degree of diversity between 0 and 1 across its two member elements. The total number of active element pairs yielded by a parent minimal cut set is defined as S_a.

(e) Classification and degree of diversity

In order to derive an overall degree of diversity from a parent minimal cut set it is appropriate to apply classifications to the individual elements of the set. The overall degree of diversity is given by the sum of the individual diversities of the active element pairs expressed as a fraction of the number of element pairs in the parent cut set. The following examples illustrate the procedure and are referred to a minimal cut set A_1A_2BC.

100% diversity
A, B and C are 100% diverse classes. Since A_1 is identical to A_2 the active elements are given by the group set ABC.

The number of active element pairs S_a is given by

$$S_a = \frac{3(3-1)}{2}$$

$$= 3$$

i.e. there are three degrees of diversity in the parent minimal cut set.

Partial diversities
Let diversity between classes A and B be partial at 75%, between A and C be 25% and between B and C be 50%.

$$S_a = 3$$

Total degrees of diversity d in parent minimal cut set are given by

$$d = (AB = 0.75) + (AC = 0.25) + (BC = 0.50)$$
$$= 1.50$$

(f) Percentage diversity D_p

This is given by the degrees of diversity d yielded by the active element pairs expressed as a percentage of the maximum possible degrees of diversity which would be generated by the parent minimal cut set if all the elements were 100% mutually diverse, i.e.

$$D_p = \frac{d}{S_p} \times 100\%$$

where S_p is the total number of element pairs in the parent minimal cut set. Referring to the following examples.

Cut set A_1A_2BC
Assume 100% diversity between classes.

$$S_p = \frac{4(4-1)}{2} = 6$$

Active element set $= ABC$

since $A_1 = A_2$.

$$S_a = \frac{3(3-1)}{2} = 3$$

$$d = (AB = 1.0) + (AC = 1.0) + (BC = 1.0) = 3$$

$$D_p = \frac{d}{S_p} \times 100\%$$

$$= \frac{3}{6} \times 100\%$$

$$= 50\%$$

Cut set $A_1A_2BC_1C_2D$
Assume 75% diversity between classes A and B, 50% between C and D and 100% between all other classes.

$$S_p = \frac{6(6-1)}{2} = 15$$

Active element set $= ABCD$

$$S_a = \frac{4(4-1)}{2} = 6$$

$$d = (AB = 0.75) + (AC = 1.0) + (AD = 1.0) + (BC = 1.0)$$
$$+ (BD = 1.0) + (CD = 0.5)$$
$$= 5.25$$

$$D_p = \frac{d}{S_p} \times 100\%$$

$$= \frac{5.25}{15} \times 100\%$$

$$= 35\%$$

Table 10.3 compiles β values for a five-element parent minimal cut set P_1, P_2, P_3, P_4 and P_5 covering a range of classes A, B, C, D and E. The table is based entirely on equipment type. It is assumed for illustrative purposes that the degree of diversity between classes A and B is 0.5, between C and D is 0.75 and between other classes is 1.00.

The relationship of β to D_p is given in Figure 10.8 and shows that the most significant improvement in β occurs between 0% and 5%. At higher diversity values above 50% the improvement in β is much less significant and subject to the law of diminishing returns in terms of higher costs associated with increasing diversification.

Table 10.3 Variable diversity independent cut sets/β-value derivations

	Cut-set elements					Degrees of diversity	Cut-set pairs S_p	Active pairs S_a	Percentage diversity D_p	β
	P_1	P_2	P_3	P_4	P_5					
Class	A	B	C	D	E	9.25	10	10	92.5	0.001
	A	A	B	C	D	5.25	10	6	52.5	0.009
	A	A	B	C	C	2.50	10	3	25	0.032
	A	A	C	C	C	1	10	1	10	0.063
	A	A	A	A	C	1	10	1	10	0.063
	A	A	A	A	A	0	10	0	0	0.100

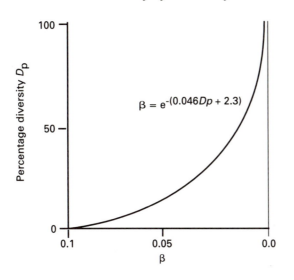

$$\beta = e^{-(0.046Dp + 2.3)}$$

Figure 10.8 Plot of β versus percentage diversity $D_{\rm p}$.

The beta–diversity relationship has not been fully endorsed by empirical data observations in the processing field. However, the law on which this methodology is based is known to be of an exponential nature in that the ratio factor $\beta = 0.1$ has been endorsed by field data where diversities are not clearly evident. At the lower end of the scale where $\beta = 0.001$, diversity is known to approach 100%. In reality, 100% diversity between redundant subsystems is difficult to achieve in practical terms since there is always a possibility that unidentified rare failure event commonalities are present.

10.8 ELEMENTARY SYSTEM STUDY

In a full-scale risk or reliability study the fault tree would yield a substantial number of independent minimal cut sets, each of which would require rapid analysis in terms of common-mode contributors.

The simplified case study to be demonstrated is described by two minimal cut sets namely *ADE* and *BCDE* respectively. Hence overall system failure would in Boolean terms be given by

$$ADE \text{ OR } BCDE, \quad \text{i.e. } ADE + BCDE$$

The element items in each cut set relate to equipment subsystems. The example will show the evaluations of:

- the system common-mode failure probability;
- the system hazard rate due to common-mode system failures.

Table 10.4 Subsystem data base

Item	Item type	Item class	Failure rate	Failure probability
A	Pressure switch	1	0.10	0.025
B	Pressure transducer	2	0.02	0.005
C	Level switch	3	0.006	0.0015
D	Automatic valve	4	0.50	0.125
E	Automatic valve	4	0.50	0.125

Subsystem proof test interval $\tau = 0.25$ year. Table 10.4 lists typical subsystem data.

Analysis of the two minimal cut sets in terms of their single-order common-mode sets is shown in Table 10.5. Figure 10.8 shows the relationship of β to D_p.

Table 10.5 Beta-value derivations from system independent minimal cut sets

Cut set	Elements and class					Degrees of diversity	Cut-set pairs S_p	Active pairs S_a	Percentage diversity D_p	β
	A	B	C	D	E					
ADE	1	–	–	4	4	1	3	1	33	0.02
BCDE	–	2	3	4	4	3	6	3	50	0.01

10.8.1 Analysis – probability mode

In the probability mode, all elements which make up the two independent minimal cut sets will represent subsystems of the protective system which is required to be analysed in terms of its common-mode failure probability.

(a) Cut set ADE

Common-mode set |ADE|
From Table 10.4 the most available subsystem is that of item A, where $P_A = 2.5 \times 10^{-2}$. Apply $\beta = 0.02$ for 33% diversity across ADE (based on equipment type). The common-mode failure probability for $|ADE|$ is then

$$P_A \times \beta = (2.5 \times 10^{-2})(2 \times 10^{-2})$$

$$= 5.0 \times 10^{-4}$$

Second-order common-mode set survey
Select, say common-mode set $|AD|E$. The percentage diversity across AD is 100%. Apply $\beta = 0.001$ for 100% diversity. The common-mode failure probability is then

$$(P_A \times \beta)P_E = (2.5 \times 10^{-2} \times 10^{-3})(1.25 \times 10^{-1})$$

$$= 3.13 \times 10^{-6}$$

Therefore the failure probability approximation for three second-order common-mode sets is

$$P_{CM} = 9.4 \times 10^{-6}$$

System ADE *common-mode failure probability*
The system failure probability is

$$P_0 = (5.0 \times 10^{-4}) + (9.4 \times 10^{-6})$$

$$= 5.1 \times 10^{-4}$$

Apply 25% concurrency factor to give

$$P_0 = 1.27 \times 10^{-4}$$

(b) Cut set BCDE

Common-mode set $|BCDE|$
From Table 10.4 the most available subsystem is that of item C where $P_C = 1.5 \times 10^{-3}$. Apply $\beta = 0.01$ for 50% diversity across $BCDE$ (based on equipment type). The common-mode failure probability $|BCDE|$ is then

$$P_C \times \beta = (1.5 \times 10^{-3})(1 \times 10^{-2})$$

$$= 1.5 \times 10^{-5}$$

Second-order common-mode set survey
Select, say $|CDE|B$. The percentage diversity across CDE is 33%. Apply $\beta = 0.02$ for 33% diversity to most available item C. The common-mode failure probability is then

$$(P_C \times \beta)P_B = (1.5 \times 10^{-3} \times 2 \times 10^{-2})(5.0 \times 10^{-3})$$

$$= 1.5 \times 10^{-7}$$

Therefore the common-mode failure probability approximation for four second-order common-mode sets is

$$P_{CM} = 4 \times 1.5 \times 10^{-7}$$
$$= 6.0 \times 10^{-7}$$

System BCDE *common-mode failure probability*

The system failure probability is

$$P_0 = (1.5 \times 10^{-5}) + (6 \times 10^{-7})$$
$$= 1.56 \times 10^{-5}$$

Apply 25% concurrency factor to give

$$P_0 = 3.9 \times 10^{-6}$$

(c) Overall system common-mode failure probability

This is given by the sum of the common-mode failures arising from the independent minimal cut sets *ADE* and *BCDE*:

$$P_0 = (1.27 \times 10^{-4}) + (3.9 \times 10^{-6})$$
$$= 1.3 \times 10^{-4}$$

(d) Conclusions

The common-mode failure probability for the overall system is assessed at 1.3×10^{-4}.

The common-mode failure contribution could be significant in a major hazard scenario.

The common-mode failure arising from subsystem *ADE* is assessed at 1.27×10^{-4} and represents 98% of the overall common-mode failure probability.

The common-mode failure arising from subsystem *BCDE* is assessed at 3.9×10^{-6} and represents 2% of the overall system common-mode failure probability.

The overall system common-mode failure is seen to be significant because of the lack of diversity between the two automatic process valves *D* and *E*. Identical valves, although advantageous from the points of view of maintenance and spares, are not seen to be conducive to the achievement of a lower common-mode system failure. It would therefore be recommended that the valves be obtained from more than one manufacturer. Maintainability by the use of different personnel in addition to a regime of staggered proof testing would also promote improvement of their mutual diversity to at least 0.5 which would improve the β factor ratio for $|ADE|$ from 0.02 to 0.005 and for $|BCDE|$ from 0.01 to 0.003.

The proof testing frequency of the automatic process valves D and E is recommended to be increased to six weeks from three months in order to further reduce their common-mode influence on the overall system.

10.8.2 Analysis – rate mode

Minimal cut sets based on rate are pertinent to risk quantification in that a process demand, namely a rate, is combined with safety system failure probability in order to derive a risk index in terms of annual hazard rate. Process demands are randomly distributed with time and their occurrences are therefore equally likely over the full period of safety system proof test intervals. In the risk mode analyses which follow, the rate or process demand elements are defined as A and B.

(a) Cut set ADE

Common-mode set $|ADE|$
From Table 10.4 the most available subsystem is that of item A where $\theta_A = 0.10$. Apply $\beta = 0.02$ for 33% diversity across ADE. The hazard rate due to common-mode failure is then

$$\theta_A \times \beta = 0.10 \times 0.02$$
$$= 2 \times 10^{-3}$$

Second-order common-mode set survey
The third-order minimal cut set yields three second-order common-mode sets, namely $A|DE|$, $D|AE|$ and $E|AD|$.

Set $A|DE|$ The percentage diversity across DE is 0, therefore $\beta = 0.1$. The hazard rate due to common-mode failure is then

$$\text{(Demand rate } \theta_A) \times \text{(Common-mode FDT across } DE)$$
$$= \theta_A \times (\mu_D \text{ or } \mu_E \times \beta)$$
$$= 0.1 \times (6.25 \times 10^{-2} \times 10^{-1})$$
$$= 6.25 \times 10^{-4}$$

Set $D|AE|$ The percentage diversity across AE is 100%, therefore $\beta = 0.001$. The hazard rate due to common-mode failure is then

$$\text{FDT}(\mu_D) \times (\theta_A \times \beta) = (6.25 \times 10^{-2}) \times (2.5 \times 10^{-2} \times 10^{-3})$$
$$= 1.56 \times 10^{-6}$$

It should be noted that $\theta_A\beta$ would be representative of the frequency of a common-cause failure condition which would at the same time negate the protection afforded by protective channel E and initiate a process demand θ_A.

This would in practice be a rare event which could typically be attributed to the onset of a general site emergency state, e.g. fire, explosion, aircraft impact etc.

Set $E|AD|$ This is identical to that discussed above for set $D|AE|$ since D and E are identical items. Hence the hazard rate due to common-mode failure is 1.56×10^{-6}.

System ADE – hazard rate due to common-mode failures
This is given by

$$|ADE| + A|DE| + D|AE| + E|AD|$$

$$= (2 \times 10^{-3}) + (6.25 \times 10^{-4}) + 2(1.56 \times 10^{-6})$$

$$= 2.63 \times 10^{-3}$$

Apply 25% concurrency factor to give

$$ADE \text{ system hazard rate} = 6.57 \times 10^{-4}$$

(b) Cut set BCDE

Common-mode set $|BCDE|$
From Table 10.4 the most available subsystem is that of item C where $\theta_C = 6 \times 10^{-3}$. Apply $\beta = 0.01$ for 50% diversity across $BCDE$. The hazard rate due to common-mode failure is then

$$\theta_C \times \beta = (6 \times 10^{-3}) \times (1 \times 10^{-2})$$

$$= 6 \times 10^{-5}$$

Second-order common-mode set survey
The fourth-order minimal cut set yields four second-order common-mode sets, namely $B|CDE|$, $C|BDE|$, $D|BCE|$ and $E|BCD|$.

Set $B|CDE|$ The percentage diversity across CDE is 33%, therefore $\beta = 0.02$. The hazard rate due to common-mode failure is then

(Demand rate θ_B) $\times$ (Common-mode FDT across $|CDE|$)

$$= \theta_B \times (\mu_C \times \beta)$$

$$= (2 \times 10^{-2}) \times (7.5 \times 10^{-4} \times 2 \times 10^{-2})$$

$$= 3.0 \times 10^{-7}$$

Set $C|BDE|$ The percentage diversity across BDE is 33%, therefore $\beta = 0.02$. The hazard rate due to common-mode failure is then

$$\text{FDT}(\mu_C) \times (\theta_B \times \beta) = (7.5 \times 10^{-4}) \times (2 \times 10^{-2} \times 2 \times 10^{-2})$$

$$= 3.0 \times 10^{-7}$$

Set $D|BCE|$ The percentage diversity across BCE is 100%, therefore $\beta = 0.001$. The hazard rate due to common-mode failure is then

$$\text{FDT}(\mu_D) \times (\theta_C \times \beta) = (6.25 \times 10^{-2}) \times (6 \times 10^{-3} \times 1 \times 10^{-3})$$

$$= 3.75 \times 10^{-7}$$

Set $E|BCD|$ This is similar to that of set $D|BCE|$ since D and E are identical items. Hence hazard rate due to common-mode failure for set $E|BCD|$ is 3.75×10^{-7}.

System BCDE – *hazard rate due to common-mode failure*
This is given by

$$|BCDE| + B|CDE| + C|BDE| + D|BCE| + E|BCD|$$

$$= (6 \times 10^{-5}) + 2(3.0 \times 10^{-7}) + 2(3.75 \times 10^{-7})$$

$$= 6.14 \times 10^{-5}$$

Apply 25% concurrency factor to give

$$\text{BCDE system hazard rate} = 1.53 \times 10^{-5}$$

(c) Overall system hazard rate due to common-mode failure

This is given by the sum of the hazard rates arising from the independent minimal cut sets ADE and $BCDE$:

$$H_0 = (6.57 \times 10^{-4}) + (1.53 \times 10^{-5})$$

$$= 6.72 \times 10^{-4}$$

(d) Conclusions

The overall system hazard rate due to common-mode failures is assessed at 6.72×10^{-4} per annum.

The system given by *ADE* contributes 98% of the system hazard rate.

The overall system is capable of improvement by diversification of the automatic process valves including a more frequent proof-testing regime. Staggered testing is also recommended in order to improve diversity.

10.8.3 Conclusions – elementary study

The study has demonstrated a methodology which enables overall system analysis of common-mode failure in terms of either probability or hazard rate.

It has been shown how each minimal cut set generated by a fault tree contributes to the overall system common-mode failure. The individual cut-set contributors tend to be, in the majority of cases, relatively rare events but can aggregate to significant values if system minimal cut sets should be unnecessarily overabundant.

Designers in their quest for higher reliabilities tend to overdesign protection on the basis of independent failure mode and achieve high levels of safety in numerical terms which become hardly credible. For example, an independent failure probability (not to be confused with hazard rate) of 10^{-5} or less based on excessive or unnecessary redundancies ceases to be credible because of the consequential rising influence of system common-mode failure which is not normally taken into account.

Defences against common-mode failure are usually met by rule-of-thumb methods where it is normal to add on redundancy without precise knowledge of where the significant common-mode influences may be in the overall system. This is not seen as a cost-effective approach, nor does it identify the specific common-mode problem areas. The methodology demonstrated overcomes these disadvantages in that it represents a mechanism whereby weak areas can be easily identified and defences confined to such areas in the overall system.

The β factor method based on fault-tree minimal cut-set analysis enables identification of common-mode influences down to levels of individual items of equipment, with considerable cost saving benefits accompanied by a greater assurance of meeting required safety criteria.

Appendices

Tutorial projects

A

Qualitative assessment of safety system reliability

A.1 PROJECT DESCRIPTION

The project is based on a protective system presented in failure mode by an engineering-level fault tree given by Figure A.1. The fault tree requires reduction in Boolean terms in order to construct the mathematical model which is shown in Figure A.2.

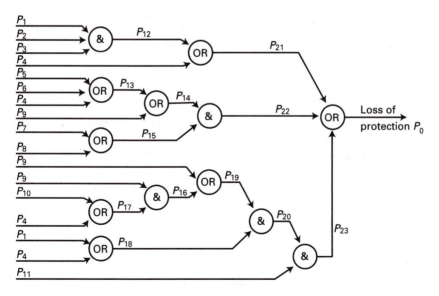

Figure A.1 Fault tree for protective system.

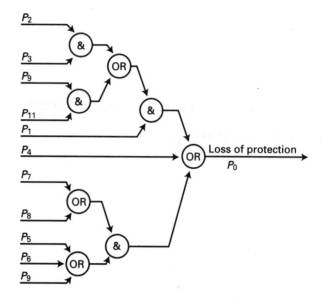

Figure A.2 Mathematical model for protective system.

The primary events are listed from P_1 to P_{11} inclusive and are used in the Boolean expressions which derive the cut-set listing from which the mathematical model is constructed. The mathematical model is probabilistic in form and is capable of accepting a data set to evaluate loss of protection in terms of system failure probability expressed either as a probability P_0 or mean fractional dead time μ_0.

A.2 PROJECT AIMS

- Produce a minimal cut-set listing by treating each primary event as a Boolean deterministic quantity.
- Produce a factorized expression for system failure probability P_0 from the minimal cut sets.
- Produce a mathematical model for loss of protection P_0.
- Analyse in qualitative terms the weak areas of the system.

A.3 MINIMAL CUT-SET LISTING

Referring to the fault tree in Figure A.1, we express P_{21}, P_{22} and P_{23} as follows:

$$P_{21} = P_4 + P_{12}$$
$$= P_4 + P_1 P_2 P_3$$

$$P_{22} = P_{14} P_{15}$$
$$= (P_9 + P_{13})(P_7 + P_8)$$
$$= (P_9 + P_4 + P_5 + P_6)(P_7 + P_8)$$
$$= P_7 P_9 + P_4 P_7 + P_5 P_7 + P_6 P_7 + P_8 P_9 + P_4 P_8 + P_5 P_8 + P_6 P_8$$

$$P_{23} = P_{11} P_{20}$$
$$= P_{11}(P_{18} P_{19})$$
$$= P_{11}[(P_1 + P_4)(P_9 + P_{16})]$$
$$= P_{11}[(P_1 + P_4)(P_9 + P_9 P_{17})]$$
$$= P_{11}[(P_1 + P_4)(P_9)(1 + P_{17})]$$
$$= P_{11}[(P_1 + P_4)P_9]$$
$$= P_1 P_9 P_{11} + P_4 P_9 P_{11}$$

The three inputs to the output OR gate as shown in the mathematical model of Figure A.2 are now combined to give the overall probability expression P_0 for the system and the minimal cut sets derived as follows:

$$P_0 = P_{21} + P_{22} + P_{23}$$
$$= P_4 + P_1 P_2 P_3 + P_7 P_9 + P_4 P_7 + P_5 P_7 + P_6 P_7 + P_8 P_9 + P_4 P_8$$
$$+ P_5 P_8 + P_6 P_8 + P_1 P_9 P_{11} + P_4 P_9 P_{11}$$

The minimal cut-set listing for P_0 is thus

$$P_0 = P_4 + P_1 P_2 P_3 + P_7 P_9 + P_5 P_7 + P_6 P_7 + P_8 P_9 + P_5 P_8 + P_6 P_8$$
$$+ P_1 P_9 P_{11}$$

A.4 FACTORIZED EXPRESSION FOR P_0

From the minimal cut-set listing,

$$P_0 = P_4 + P_1 P_2 P_3 + P_7(P_5 + P_6 + P_9) + P_8(P_5 + P_6 + P_9)$$
$$+ P_1 P_9 P_{11}$$

The factorized expression is thus

$$P_0 = P_4 + P_1(P_2 P_3 + P_9 P_{11}) + (P_7 + P_8)(P_5 + P_6 + P_9)$$

A.5 MATHEMATICAL MODEL

From the factorized expression P_0 the mathematical model given in Figure A.2 is constructed. The model shows all the primary input failure events on the left-hand side and their logical progression to the end event, namely loss of protection, shown on the right-hand side. Comparison with the fault tree of Figure A.1 shows little apparent similarity in form. In fact the two are equivalent, with the difference being that the fault tree is not Boolean reduced and therefore contains redundant elements. Reduction effectively reconstructs the fault tree into a reduced form which is called the mathematical model. In order to demonstrate equivalence the reader is recommended to apply one or more minimal cut sets, chosen at random, to both the fault tree and mathematical model and hence to verify that the given cut set does produce the stated end event.

A.6 ANALYSIS OF THE MATHEMATICAL MODEL

In the practical situation, although the primary input events may be related to specific types of equipment or human elements, quantified data may not be available. To apply data would necessitate access to data bank sources or research into maintenance records which may not be readily compatible with data identification and therefore data access could be a relatively costly procedure. It follows that there is a monetary advantage as well as time saving in analysing a given mathematical model in qualitative terms for prominence of weak areas. This allows a later stage detailed assessment confined to weak areas when equipment and design details can be more closely specified.

The mathematical model which describes this particular project is put forward without any details of the representative data set. However, the proposed equipments related to the primary inputs will be normally known in principle and hence selective judgements made on their prominence in the model.

In this particular case it will be observed that the topmost input to the final OR gate of the mathematical model in Figure A.2 will most likely be less significant than the other two. It will be seen that the particular input relies on transition through three AND gates and since fractional quantities are present, this uppermost input will most likely be relatively smaller in magnitude than the other two. Hence in qualitative terms, it is concluded that the lowermost inputs will represent the more significant areas of the system under consideration.

A.7 RECOMMENDATIONS

Consideration should be given to invest redundancy and diversity into the equipment related to input P_4. This recommendation is made since P_4 is a single input to the final OR gate and is likely to be significant to the overall system failure. Diversity is also called for in order to reduce the prominence of common-mode failure which would arise across the redundant channels.

Redundancy with possible diversity should be considered for the anticipated higher failure rate input of the OR gate set P_7 and P_8.

Redundancy with possible diversity should be considered for the anticipated highest failure rate input of the OR gate set P_5, P_6 and P_9.

The mathematical model comprises eight multi-element minimal cut sets which are each a source of system common-mode failure. The system will therefore be susceptible to common-mode failures which may need to be assessed, dependent on the nature of the hazard and degree of acceptable risk. It is recommended in the first instance that one design objective should be to ensure that the number of minimal cut sets be reduced to an optimum without impairing the safety philosophy.

B

Quantitative assessment of safety system reliability

B.1 PROJECT DESCRIPTION

Project B relates to the fault tree and derived mathematical model of Project A, namely Figures A.1 and A.2, and extends this first project to a quantified form by the application of a typical data base.

B.2 ASSESSMENT AIMS

- Assess the system independent mean fractional dead time μ_0.
- Estimate for purposes of sensitivity analysis FDT values at the output of each logic gate.
- Comment on the weak areas of the model and show that μ_4 and μ_7 inputs to the system output gate 9 are significant contributors to the output value P_0.
- Make recommendations on how to improve the system independent failure FDT μ_0 in order to be compatible with a criterion value of 10^{-3}.
- Assess the overall enhanced system in respect of common-mode mean fractional dead time consequent to the implementation of those recommendations in respect of meeting a compatible value of independent failure FDT.
- Comment on the influence of common-mode failures on the overall enhanced system. Identify the location of significant failures.

B.3 ASSESSMENT TERMS OF REFERENCE

- Proof test interval = 0.25 year.

- Failure rates are in terms of faults per annum.
- Assume 25% concurrency of failures factor for dependent failures.
- Data base is given in Table B.1.

Table B.1 Subsystem data base

Subsystem	Ref.	Failure rate
Flowmeter	P_1	0.2
Pressure switch	P_2	0.04
Relay	P_3	0.001
Oxygen analyser	P_4	0.35
Temperature sensor	P_5	0.18
Solenoid valve	P_6	0.07
Level switch	P_7	0.33
Process valve	P_8	0.5
Pressure transducer	P_9	0.016
Rotation detector	P_{10}	0.7
Trip amplifier	P_{11}	0.023

B.4 OVERALL SYSTEM INDEPENDENT FAILURE FDT μ_0

The mathematical model is given by Figure B.1 and shows how primary failure events lead to loss of system protection. The model is based on the cut-set listing of Project A.

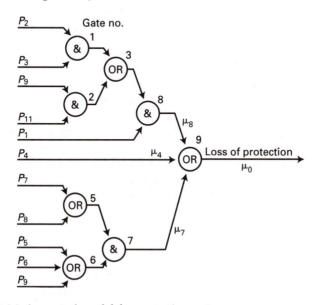

Figure B.1 Mathematical model for protective system.

Hence

$$\mu_0 = \frac{1}{\tau} \int_0^\tau P_0 \, dt$$

$$= \frac{1}{\tau} \int_0^\tau (P_4 + P_1 P_2 P_3 + P_7 P_9 + P_5 P_7 + P_6 P_7 + P_8 P_9 + P_5 P_8 + P_6 P_8$$

$$+ P_1 P_9 P_{11}) \, dt$$

Integrating and inserting values:

$$\mu_0 = \frac{0.35 \times 0.25}{2} + \frac{0.2 \times 0.04 \times 0.001 \times (0.25)^3}{4}$$

$$+ \frac{0.33 \times 0.016 \times (0.25)^2}{3} + \frac{0.18 \times 0.33 \times (0.25)^2}{3}$$

$$+ \frac{0.07 \times 0.33 \times (0.25)^2}{3} + \frac{0.5 \times 0.016 \times (0.25)^2}{3}$$

$$+ \frac{0.18 \times 0.5 \times (0.25)^2}{3} + \frac{0.07 \times 0.5 \times (0.25)^2}{3}$$

$$+ \frac{0.2 \times 0.016 \times 0.023 \times (0.25)^3}{4}$$

$$\mu_0 = 4.84 \times 10^{-2}$$

B.4.1 Cautionary note

If FDTs are low they may be added directly through OR gates. FDTs may be taken through AND gates by direct multiplication using correction factors, providing the denominators of the individual FDTs are identical. Where denominators are not identical then the use of correction factors will provide approximate gate output values. This is permissible in estimating intermediate values between the primary input events and the output of the mathematical model for purposes of sensitization analysis. The final, usually more accurate, output value of the model may be derived from the overall equation as demonstrated above.

B.5 SENSITIVITY ANALYSIS

This requires estimations of individual logic gate outputs of the mathematical model given by Figure B.1 in order to identify the weak areas of the system.

Gate 1

$$P = P_2 P_3$$

$$\mu_1 = \frac{0.04 \times 0.001 \times (0.25)^2}{3} = 8.33 \times 10^{-7}$$

Gate 2

$$P = P_9 P_{11}$$

$$\mu_2 = \frac{0.016 \times 0.023 \times (0.25)^2}{3} = 7.67 \times 10^{-6}$$

Gate 3

$$\mu_3 = \mu_1 + \mu_2$$
$$\mu_3 = 8.5 \times 10^{-6}$$

Gate 8

$$\mu_8 = \frac{4}{3} \mu_3 \frac{1}{\tau} \int_0^\tau P_1 \, dt = \frac{4}{3} \times 8.5 \times 10^{-6} \times \frac{0.2 \times 0.25}{2}$$

$$= (\ll 10^{-5})$$

Gate 5

$$P = P_7 + P_8$$

$$\mu_5 = \frac{0.33 \times 0.25}{2} + \frac{0.5 \times 0.25}{2}$$

$$\mu_5 = 1.03 \times 10^{-1}$$

Gate 6

$$P = P_5 + P_6 + P_9$$

$$\mu_6 = \frac{0.18 \times 0.25}{2} \times \frac{0.07 \times 0.25}{2} + \frac{0.016 \times 0.25}{2}$$

$$\mu_6 = 3.33 \times 10^{-2}$$

Gate 7

$$\mu_7 = \mu_5 \times \mu_6 \times \frac{4}{3}$$

$$\mu_7 = 4.57 \times 10^{-3}$$

Gate 9

$$\mu_0 = \mu_8 + \mu_4 + \mu_7$$

$$= (\ll 10^{-5}) + \frac{0.35 \times 0.25}{2} + 4.57 \times 10^{-3}$$

$$= (\ll 10^{-5}) + 4.38 \times 10^{-2} + 4.57 \times 10^{-3}$$

$$\mu_0 = 4.84 \times 10^{-2}$$

B.5.1 Conclusions and comments

The assessed value of μ_0 for the system independent failure FDT shows that it is incompatible with the 10^{-3} criterion. The following conclusions and comments are therefore given in order to promote those improvements which would be conducive to the criterion objective. It should be noted that the conclusions and comments are made with due allowances for common-mode or dependent failures which will need to be anticipated when specifying an enhanced system.

The assessment shows that:

- the significant subsystem independent failure probabilities given by μ_4 and μ_7 are 4.38×10^{-2} and 4.57×10^{-3} respectively;
- redundancy is lacking in the oxygen analyser channel P_4;
- redundancy would be desirable in the subsystem P_8 input to gate 5;
- redundancy would be desirable in the subsystem P_5 input to gate 6.

B.6 RECOMMENDATIONS

It is recommended that the following be modified into redundant logic compatible with an overall system FDT criterion of 10^{-3}:

- the analyser channel P_4;
- the process shutdown valve system P_8;
- the temperature sensor channel P_5.

B.7 IMPLEMENTING RECOMMENDATIONS

B.7.1 Oxygen analyser channel P_4

Simple 'one-out-of-three' redundant success voting logic equates to an FDT of 1.67×10^{-4}.

B.7.2 Process automatic shutdown system P_8

Simple 'one-out-of-two' redundant success voting logic equates to an FDT of 5.2×10^{-3}.

B.7.3 Temperature sensor channel P_5

Simple 'one-out-of-two' redundant success voting logic equates to an FDT of 6.75×10^{-4}. Note P_5 has the highest failure rate of the input set to gate 6 and hence will be the most significant means of reducing the gate output FDT if redundancy is employed.

B.8 OVERALL ENHANCED SYSTEM – INDEPENDENT FAILURE REVIEW

Revision of the original system in accordance with recommendations is depicted by the mathematical model shown in Figure B.2. The output μ_8 from gate 8 remains at a negligible value $\ll 10^{-5}$ since no changes have been necessary in this area of the original protective system model.

The output μ_4 is improved by the now redundant inputs to gate 4. Gate 5 is likewise modified by redundant inputs into gate 5A, whilst gate 6 is also modified by redundant inputs to gate 6A.

B.8.1 Enhanced system independent FDT calculation

The equation for the enhanced mathematical model is taken from Figure B.2 and neglects μ_8 since its value is $\ll 10^{-5}$. Hence

$$P_0 \text{ (effective)} = P_{4A}P_{4B}P_{4C} + (P_{8A}P_{8B} + P_7)(P_{5A}P_{5B} + P_6 + P_9)$$

$$\text{Enhanced system FDT } \mu_0 = \frac{1}{\tau} \int_0^\tau P_0 \, dt$$

In each redundant subsystem the input elements are assumed for simplicity to have identical failure rates. Hence $P_{4A}P_{4B}P_{4C} = (P_4)^3$ etc. Inserting values into P_0 and integrating:

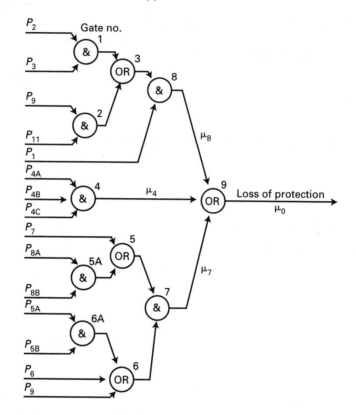

Figure B.2 Enhanced mathematical model.

$$\mu_0 = \frac{1}{\tau} \int_0^\tau 0.35^3 t^3 + (0.5^2 t^2 + 0.33t)(0.18^2 t^2 + 0.07t + 0.016t) \, dt$$

where $\tau =$ proof test interval $= 0.25$ year. Thus

$$\mu_0 = \frac{1}{\tau} \int_0^\tau 0.35^3 t^3 + 0.008t^4 + 0.0175t^3 + 0.004t^3 + 0.01t^3$$

$$+ \, 0.023t^2 + 0.005t^2 \, dt$$

$$\mu_0 = \frac{0.008 \times 0.25^4}{5} + \frac{0.25^3}{4}(0.35^3 + 0.0175 + 0.004 + 0.01)$$

$$+ \frac{0.25^2}{3}(0.023 + 0.005)$$

$$\mu_0 = 8.8 \times 10^{-4}$$

B.8.2 Conclusion

The enhanced system design is seen to be capable of achieving a mean fractional dead time in independent failure terms of 8.8×10^{-4}. This is seen to be compatible with the desired criterion providing the common-mode failure of the modified system is maintained at an acceptable level.

B.9 COMMON-MODE ANALYSIS

In order to achieve an overall system criterion of 10^{-3} it is necessary to take into account those subsystem failures which are due to common causes. Referring to the enhanced mathematical model in Figure B.2 a minimal cut-set listing is derived which includes redundancies resulting from the recommendations of section B.7.

Let P_0 be the probability equation derived from the enhanced model in Figure B.2. Then

$$P_0 = (P_2P_3 + P_9P_{11})P_1 + P_{4A}P_{4B}P_{4C} + (P_7 + P_{8A}P_{8B})(P_{5A}P_{5B} + P_6 + P_9)$$

where subscripts A, B etc. refer to the individual redundant channels. Expanding the equation gives the following minimal cut-set listing:

$$P_0 = P_1P_2P_3 + P_1P_9P_{11} + P_{4A}P_{4B}P_{4C} + P_7P_{5A}P_{5B} + P_6P_7 + P_7P_9$$
$$+ P_{5A}P_{5B}P_{8A}P_{8B} + P_6P_{8A}P_{8B} + P_9P_{8A}P_{8B}$$

B.9.1 Common-mode analysis listing

The analysis shown in Table B.2 reveals that the assessed overall system FDT due to common-mode failures is 5.8×10^{-3} of which 76% is due to the three redundant oxygen analyser channels given by ident 3.

Table B.2 Common-mode analysis listing

Ident	Cut set	D_p (%)	β	Common-mode FDT μ_{CM}
1	$P_1P_2P_3$	100	0.001	1.3×10^{-7}
2	$P_1P_9P_{11}$	100	0.001	2.0×10^{-6}
3	$P_{4A}P_{4B}P_{4C}$	0	0.1	4.4×10^{-3}
4	$P_7P_{5A}P_{5B}$	33	0.02	4.5×10^{-4}
5	P_6P_7	100	0.001	8.8×10^{-6}
6	P_7P_9	100	0.001	2.0×10^{-6}
7	$P_{5A}P_{5B}P_{8A}P_{8B}$	25	0.03	6.8×10^{-4}
8	$P_6P_{8A}P_{8B}$	33	0.02	1.8×10^{-4}
9	$P_9P_{8A}P_{8B}$	33	0.02	4.0×10^{-5}
			Total	5.8×10^{-3}

Allow 25% concurrency of failures factor; then the effective system common-mode FDT $\mu_{CM} = 1.45 \times 10^{-3}$.

B.10 REVIEW OF ENHANCED SYSTEM

Reassessment of the overall system following the recommendations of section B.7 shows that:

- the independent FDT of the system is capable of achieving 8.8×10^{-4};
- the system common-mode FDT is assessed at 1.45×10^{-3};
- the overall system FDT $\mu_0 = 8.8 \times 10^{-4} + 1.45 \times 10^{-3} = 2.33 \times 10^{-3}$;
- the system independent-mode FDT represents 38% of the overall system failure probability on demand;
- the system common-mode FDT represents 62% of the overall system failure probability on demand.

B.11 SUMMARY

The assessed overall system FDT of 2.33×10^{-3} could be described as 'worse than 10^{-3}'. It has been shown that the protective system failure probability is most susceptible to common-mode failures. It therefore follows that improvements in this mode would be most beneficial in achieving the system FDT criterion of 10^{-3}. The common-mode analysis listing at ident 3 in Table B.2 highlights the prominence of the three oxygen analyser channels which accounts for 75% of the overall system common-mode failure. A means of reducing the common-mode contribution from the analysers would be to invest partial diversity in the three channels to the extent of 30% or more and would in consequence realize the benefit of a reduction in the related β factor from 0.1 to 0.025 or less. This measure would improve the common-mode FDT contribution from the analyser cut set to at least 1.1×10^{-3}, which equates to 2.75×10^{-4} with the 25% concurrency of failures factor taken into account.

The system overall common-mode FDT would therefore be improved from 1.45×10^{-3} to at least 6.25×10^{-4}.

Partial diversity in the oxygen analyser protective channels of the order of 30% to 50% could be achieved by:

- choosing a second manufacturer for one analyser by, say P_{4A};
- employing a different principle of measurement for P_{4A};
- applying two separate maintenance regimes to the two analyser types;
- staggered proof testing of the three channel elements.

Hence the overall enhanced system FDT μ_0 would be:

Independent failure FDT + Common-mode failure FDT

$$= (8.8 \times 10^{-4}) + (6.25 \times 10^{-4})$$

$$\mu_0 = 1.51 \times 10^{-3} \text{ with 30\% oxygen channel diversity}$$

$$\mu_0 = 1.34 \times 10^{-3} \text{ with 50\% oxygen channel diversity}$$

The practical problem of investing diversity in the oxygen protective channels lies in obtaining equipment with approximately similar failure rates and maintaining two sets of spares. An alternative solution would be to proof test the analysers at a greater frequency, say monthly or less. This would lead to benefits in both the independent and dependent failure probability modes but would entail higher maintenance overheads on the plant.

B.12 CONCLUDING REMARKS

The project is intended to demonstrate a practical technique of quantitative safety and reliability assessment when applied to process safety systems. The technique demonstrated is equally applicable to any safety system whether it be security, fire protection, alarms, electrical plant, performance of domestic equipment and in fact any system for which failure probability on demand is a requirement.

The project illustrates the importance of determining both the independent and dependent failures which result in the overall system failure probability on demand. It also shows how weak areas in both modes of failure can be identified so as to enable cost effective measures to be taken either at the conceptual design stage or the later operational stage of the plant.

At the early design stage when detailed specifications of equipment are rarely known, generic data sets may be applied and weak areas of the design concepts identified.

For operational plants where potentially costly overall quantitative assessments may be under consideration, a cheaper alternative may be possible in that an established mathematical model can be analysed in qualitative rather than quantitative terms. Specific weak areas can then be developed into quantitative form with obvious savings in detailed assessment cost.

C

Reliability case study of an automatic protective system

C.1 INTRODUCTION

Project C, shown in Figure C.1, relates to an automatic protective system based on a real-life major hazard scenario in the chemical industry. It is therefore fully endorsed as an example of a practical application and demonstration of assessment methods. The technique, supported by a representative data base, progresses from the system description through fault-tree construction to mathematical modelling. The process of minimal cut-set listing, which is dealt with in substantial detail, precedes the independent and dependent failure modelling from which conclusions followed by recommendations are made towards meeting a desired criterion of failure probability.

C.2 PROJECT DESCRIPTION

The project relates to an automatic protective system shown in Figure C.1.

The safety system comprises three subsystems, namely, input sensor, interposing logic, and shutdown systems. The input system features diversity in terms of flow and temperature given by F and T respectively. The output from the flow sensor is connected into the input of a logic system L1, whilst a second logic system L2 receives the output from the temperature sensor. Outputs from these two logic systems are also arranged via interposing relays R1 and R2 to provide inputs to a third common logic system L3.

The flow sensor channel is arranged to directly trip solenoid valve SV1 via logic module L1. Additionally, the output from L1 is also

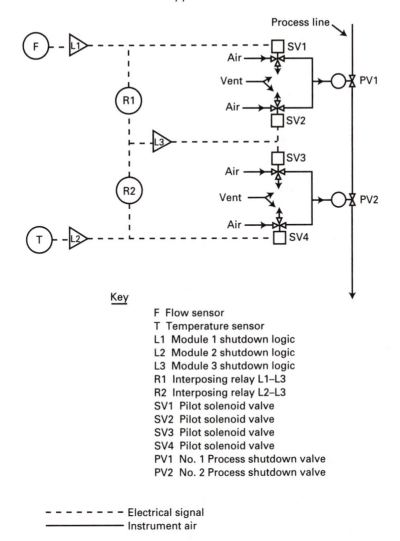

Key

F Flow sensor
T Temperature sensor
L1 Module 1 shutdown logic
L2 Module 2 shutdown logic
L3 Module 3 shutdown logic
R1 Interposing relay L1–L3
R2 Interposing relay L2–L3
SV1 Pilot solenoid valve
SV2 Pilot solenoid valve
SV3 Pilot solenoid valve
SV4 Pilot solenoid valve
PV1 No. 1 Process shutdown valve
PV2 No. 2 Process shutdown valve

– – – – – – – Electrical signal
——————— Instrument air

Figure C.1 Shutdown system schematic.

taken via relay R1 and logic module L3 to trip solenoid valves SV2 and SV3. Similarly the temperature sensor channel output is taken directly via logic module L2 to directly trip solenoid valve SV4, but additionally via relay R2 and logic module L3, the output from L2 is also arranged to trip solenoid valves SV2 and SV3.

The four solenoid valves are arranged such that either SV1 or SV2 can vent the diaphragm actuator of process valve PV1 to give process line closure and SV3 or SV4 can likewise vent the actuator of a second process valve PV2. Successful automatic process shutdown is deemed to have been achieved if either PV1 or PV2 closes on demand.

C.3 PROJECT AIMS

The overall assessment will provide:

- an engineering level fault tree;
- a minimal cut-set listing;
- a quantified mathematical model comprising
    (a) system independent failure probability on demand,
    (b) system common-mode failure probability on demand,
    (c) system overall failure probability on demand,
    (d) recommendations to achieve a system FDT of 10^{-4}.

C.4 ASSESSMENT TERMS OF REFERENCE

- System schematic (Figure C.1);
- system fault tree (Figure C.2);
- matrix solution for system failure probability P_0 (Table C.1);
- mathematical model (Figure C.3).

C.5 ASSESSMENT APPROACH

Referring to the fault tree in Figure C.2, the Boolean equation describing the overall safety system is given by:

$$P_0 = P_1 P_{24}$$

Prior to multiplication, each of the equations P_1 and P_{24} are firstly expressed by their respective minimal cut-set listings. The most convenient method of deriving the product $P_1 P_{24}$ is to multiply each of their minimal cut-set listings by means of a matrix as shown in Table C.1. The matrix shows only the minimal cut sets which constitute the overall system P_0. The blank entries contain non-minimal cut sets and are therefore superfluous to the assessment. In other words, for clarity, any non-minimal cut set contains a specific minimal cut set given in the matrix. The reader is strongly advised to construct or systematically verify the matrix in order to gain understanding and familiarity with the process of identification of minimal and non-minimal cut sets.

From the assembly of minimal cut sets enabled by the matrix the overall system equation P_0 is finally expressed by factorization of the sets from which the mathematical model is constructed.

Table C.1 Matrix solution for $P_0 = P_1 P_{24}$

$$P_1 = P_2 + (P_5 + P_7 + P_8)(P_{10} + P_{12}) + (P_{18} + P_{20} + P_{21})(P_7 + P_8 + P_5 P_{15})$$

P_{24}	P_2	$P_5 P_{10}$	$P_7 P_{10}$	$P_8 P_{10}$	$P_5 P_{12}$	$P_7 P_{12}$	$P_8 P_{12}$	$P_7 P_{18}$	$P_7 P_{20}$	$P_7 P_{21}$	$P_8 P_{18}$	$P_8 P_{20}$	$P_8 P_{21}$	$P_5 P_{15} P_{18}$	$P_5 P_{15} P_{20}$	$P_5 P_{15} P_{21}$
P_{25}	$P_2 P_{25}$	$P_5 P_{10} P_{25}$	$P_7 P_{10} P_{25}$	$P_8 P_{10} P_{25}$	$P_5 P_{12} P_{25}$	$P_7 P_{12} P_{25}$	$P_8 P_{12} P_{25}$	$P_7 P_{18} P_{25}$	–	–	$P_8 P_{18} P_{25}$	–	–	$P_5 P_{15} P_{18} P_{25}$	–	–
$P_{12} P_{20}$	$P_2 P_{12} P_{20}$	–	–	–	$P_5 P_{12} P_{20}$	–	–	–	–	–	–	–	–	–	–	–
$P_{12} P_{21}$	$P_2 P_{12} P_{21}$	–	–	–	$P_5 P_{12} P_{21}$	–	–	–	–	–	–	–	–	–	–	–
$P_{12} P_{30}$	$P_2 P_{12} P_{30}$	–	–	–	$P_5 P_{12} P_{30}$	$P_7 P_{12} P_{30}$	$P_8 P_{12} P_{30}$	–	–	–	–	–	–	–	–	–
$P_{20} P_{29}$	$P_2 P_{20} P_{29}$	$P_5 P_{10} P_{20} P_{29}$	–	–	–	–	–	–	–	–	–	–	–	–	–	–
$P_{21} P_{29}$	$P_2 P_{21} P_{29}$	$P_5 P_{10} P_{21} P_{29}$	–	–	–	–	–	–	–	–	–	–	–	–	–	–
$P_{29} P_{30}$	$P_2 P_{29} P_{30}$	$P_5 P_{10} P_{29} P_{30}$	$P_7 P_{10} P_{29} P_{30}$	$P_8 P_{10} P_{29} P_{30}$	–	–	–	–	–	–	–	–	–	–	–	–
$P_7 P_{20}$	–	–	–	–	–	–	–	–	$P_7 P_{20}$	–	–	–	–	–	–	–
$P_8 P_{20}$	–	–	–	–	–	–	–	–	–	–	–	$P_8 P_{20}$	–	–	–	–
$P_{15} P_{20}$	$P_2 P_{15} P_{20}$	–	–	–	–	–	–	–	–	–	–	–	–	–	$P_5 P_{15} P_{20}$	–
$P_7 P_{21}$	–	–	–	–	–	–	–	–	–	$P_7 P_{21}$	–	–	–	–	–	–
$P_8 P_{21}$	–	–	–	–	–	–	–	–	–	–	–	–	$P_8 P_{21}$	–	–	–
$P_{15} P_{21}$	$P_2 P_{15} P_{21}$	–	–	–	–	–	–	–	–	–	–	–	–	–	–	$P_5 P_{15} P_{21}$
$P_7 P_{18} P_{30}$	–	–	–	–	–	–	–	$P_7 P_{18} P_{30}$	–	–	–	–	–	–	–	–
$P_8 P_{18} P_{30}$	–	–	–	–	–	–	–	–	–	–	$P_8 P_{18} P_{30}$	–	–	–	–	–
$P_{15} P_{18} P_{30}$	$P_2 P_{15} P_{18} P_{30}$	–	–	–	–	–	–	–	–	–	–	–	–	$P_5 P_{15} P_{18} P_{30}$	–	–

$$P_{24} = P_{25} + (P_{20} + P_{21} + P_{30})(P_{12} + P_{29}) + (P_7 + P_8 + P_{15})(P_{20} + P_{21} + P_{18} P_{30})$$

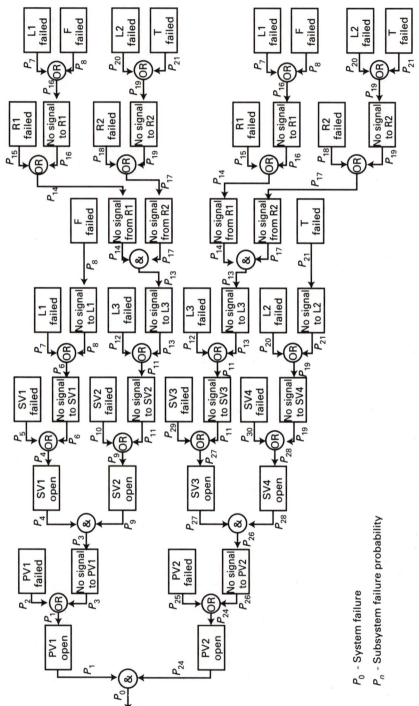

Figure C.2 Engineering fault-tree logic for shutdown system failure.

P_0 - System failure

P_n - Subsystem failure probability

Ref.	Component	Annual failure rate θ	Proof test interval (years)	FDT	Fault-tree event
PV1	No.1 Process valve	0.21	0.25	2.6×10^{-2}	P_2
PV2	No.2 Process valve	0.21		2.6×10^{-2}	P_{25}
L2	No.2 Logic module	0.003		3.8×10^{-4}	P_{20}
T	Temp. channel	0.15		1.9×10^{-2}	P_{21}
L3	No.3 Logic module	0.003		3.8×10^{-4}	P_{12}
SV2	Solenoid valve	0.01		1.3×10^{-3}	P_{10}
SV3	Solenoid valve	0.01		1.3×10^{-3}	P_{29}
SV1	Solenoid valve	0.01		1.3×10^{-3}	P_5
L1	No.1 Logic module	0.003		3.8×10^{-4}	P_7
F	Flowmeter	0.11		1.4×10^{-2}	P_8
SV1	Solenoid valve	0.01		1.3×10^{-3}	P_5
L1	No.1 Logic module	0.003		3.8×10^{-4}	P_7
F	Flowmeter	0.11		1.4×10^{-2}	P_8
PV2	No.2 Process valve	0.21		2.6×10^{-2}	P_{25}
SV4	Solenoid valve	0.01		1.3×10^{-3}	P_{30}
L3	No.3 Logic module	0.003		3.8×10^{-4}	P_{12}
SV3	Solenoid valve	0.01		1.3×10^{-3}	P_{29}
SV4	Solenoid valve	0.01		1.3×10^{-3}	P_{30}
PV2	No.2 Process valve	0.21		2.6×10^{-2}	P_{25}
SV2	Solenoid valve	0.01		1.3×10^{-3}	P_{10}
L2	No.2 Logic module	0.003		3.8×10^{-4}	P_{20}
T	Resist thermometer	0.15		1.9×10^{-2}	P_{21}
R2	Relay	0.008		1×10^{-3}	P_{18}
SV4	Solenoid valve	0.01		1.3×10^{-3}	P_{30}
PV1	No.1 Process valve	0.21		2.6×10^{-2}	P_2
SV1	Solenoid valve	0.01		1.3×10^{-3}	P_5
R1	Relay	0.008		1×10^{-3}	P_{15}
L2	No.2 Logic module	0.003		3.8×10^{-4}	P_{20}
T	Resist thermometer	0.15		1.9×10^{-2}	P_{21}
SV4	Solenoid valve	0.01		1.3×10^{-3}	P_{30}
L3	No.3 Logic module	0.003		3.8×10^{-4}	P_{12}
SV3	Solenoid valve	0.01		1.3×10^{-3}	P_{29}
PV1	No.1 Process valve	0.21		2.6×10^{-2}	P_2
L1	No.1 Logic module	0.003		3.8×10^{-4}	P_7
F	Flowmeter	0.11		1.4×10^{-2}	P_8
R2	Relay	0.008		1×10^{-3}	P_{18}
SV4	Solenoid valve	0.01		1.3×10^{-3}	P_{30}
L1	No.1 Logic module	0.003		3.8×10^{-4}	P_7
F	Flowmeter	0.11		1.4×10^{-2}	P_8
SV1	Solenoid valve	0.01		1.3×10^{-3}	P_5
R1	Relay	0.008		1×10^{-3}	P_{15}
R2	Relay	0.008		1×10^{-3}	P_{18}
PV2	No.2 Process valve	0.21		2.6×10^{-2}	P_{25}
	Common mode			7.93×10^{-4}	μ_{CM}

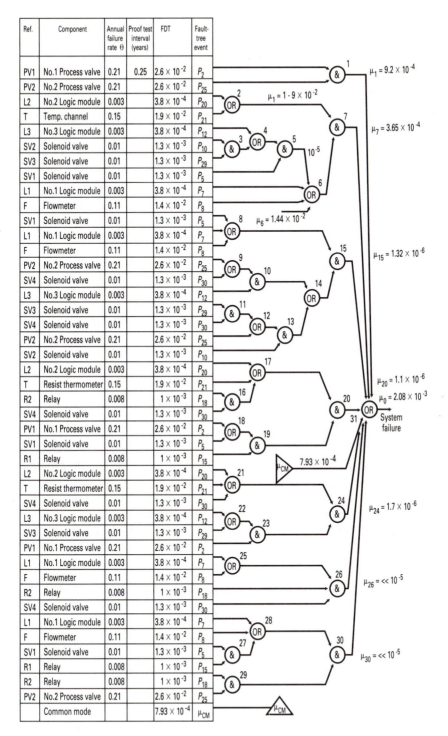

Figure C.3 Mathematical model.

C.6 BASIS OF THE ASSESSMENT

C.6.1 Data base

Table C.2 lists representative dangerous unrevealed failure rates for the subsystems which make up the overall safety system, given in terms of faults per annum.

Table C.2 Subsystem data base

Subsystem	Ref.	Failure rate
Flowmeter	F	0.11
Temperature sensor	T	0.15
Logic	L	0.003
Relay	R	0.008
Solenoid valve	SV	0.01
Process valve	PV	0.214

C.7 BOOLEAN REDUCTION OF THE FAULT TREE

Reduction of the fault tree is most conveniently carried out in three specific stages, as follows:

- From the fault tree in Figure C.2 derive a reduced equation for P_1 – failure of process valve PV1 to open on demand.
- From the fault tree in Figure C.2 derive a reduced equation for P_{24} – failure of process valve PV2 to open on demand.
- By means of the matrix in Table C.1 derive the product of $P_1 P_{24}$ in terms of minimal cut sets.

Evaluate P_1:

$$P_1 = P_2 + P_3 = P_2 + P_4 P_9 = P_2 + (P_5 + P_6)(P_{10} + P_{11})$$

$$= P_2 + (P_5 + P_7 + P_8)(P_{10} + P_{12} + P_{13})$$

$$= P_2 + (P_5 + P_7 + P_8)(P_{10} + P_{12} + P_{14}P_{17})$$

$$= P_2 + (P_5 + P_7 + P_8)[P_{10} + P_{12} + (P_{15} + P_7 + P_8)(P_{18} + P_{20} + P_{21})]$$

$$= P_2 + (P_5 + P_7 + P_8)(P_{10} + P_{12} + P_{15}P_{18} + P_{15}P_{20} + P_{15}P_{21}$$

$$+ P_7 P_{18} + P_7 P_{20} + P_7 P_{21} + P_8 P_{18} + P_8 P_{20} + P_8 P_{21})$$

$$P_1 = P_2 + (P_5 + P_7 + P_8)(P_{10} + P_{12}) + (P_{18} + P_{20} + P_{21})(P_7 + P_8 + P_5 P_{15})$$

Evaluate P_{24}:

$$P_{24} = P_{25} + P_{26} = P_{25} + P_{27}P_{28} = P_{25} + (P_{29} + P_{11})(P_{30} + P_{19})$$

$$= P_{25} + (P_{29} + P_{12} + P_{13})(P_{30} + P_{20} + P_{21})$$

$$= P_{25} + (P_{29} + P_{12} + P_{14}P_{17})(P_{30} + P_{20} + P_{21})$$

$$= P_{25} + (P_{30} + P_{20} + P_{21})[P_{29} + P_{12} + (P_{15} + P_7 + P_8)(P_{18} + P_{20} + P_{21})]$$

$$= P_{25} + (P_{30} + P_{20} + P_{21})(P_{29} + P_{12} + P_{15}P_{18} + P_{15}P_{20} + P_{15}P_{21}$$

$$+ P_7P_{18} + P_7P_{20} + P_7P_{21} + P_8P_{18} + P_8P_{20} + P_8P_{21})$$

$$P_{24} = P_{25} + (P_{20} + P_{21} + P_{30})(P_{12} + P_{29})$$

$$+ (P_7 + P_8 + P_{15})(P_{20} + P_{21} + P_{18}P_{30})$$

Evaluate P_0: from the Boolean reduction matrix,

$$P_0 = P_1P_{24}$$

$$= P_5P_{10}P_{29}(P_{20} + P_{21}) + P_{10}P_{29}P_{30}(P_5 + P_7 + P_8)$$

$$+ P_5P_{12}(P_{20} + P_{21}) + P_{12}P_{30}(P_5 + P_7 + P_8)$$

$$+ P_5P_{15}(P_{20} + P_{21} + P_{18}P_{30}) + P_{18}P_{30}(P_7 + P_8)$$

$$+ P_7(P_{20} + P_{21}) + P_8(P_{20} + P_{21}) + P_{10}P_{25}(P_5 + P_7 + P_8)$$

$$+ P_{12}P_{25}(P_{20} + P_{21} + P_{30}) + P_2P_{12}(P_{20} + P_{21} + P_{30})$$

$$+ P_2P_{15}(P_{20} + P_{21} + P_{18}P_{30}) + P_{25}$$

Factorizing gives the following reduced expression from which a mathematical model can now be produced.

$$P_0 = P_2P_{25} + \{(P_{20} + P_{21})[P_7 + P_8 + P_5(P_{12} + P_{10}P_{29})]\}$$

$$+ \{(P_5 + P_7 + P_8)[P_{12}(P_{25} + P_{30}) + P_{10}(P_{25} + P_{29}P_{30})]\}$$

$$+ \{(P_{20} + P_{21} + P_{18}P_{30})[P_{15}(P_2 + P_5)]\}$$

$$+ \{(P_{20} + P_{21} + P_{30})[P_2(P_{12} + P_{29})]\}$$

$$+ [P_{18}P_{30}(P_7 + P_8)] + [P_{18}P_{25}(P_7 + P_8 + P_5P_{15})]$$

C.8 SENSITIVITY ANALYSIS – INDEPENDENT FAILURE MODE

Referring to the mathematical model in Figure C.3, estimations of logic gate outputs are a valuable means of revealing sensitive areas of the mathematical model from which recommendations for improvements may be made. This procedure is not recommended as a method of determining the overall system mean fractional dead time, which is more accurately carried out by calculating each of the inputs to the final logic OR gate through integration of the individual probability expressions.

C.8.1 Terms of reference

- All logic gate outputs are calculated in terms of mean fractional dead times μ.
- Probability expressions are given in terms of P.
- Proof test interval $\tau = 0.25$ year.
- The mathematical model is shown in Figure C.3.

Gate 1

$$P = P_2 P_{25}$$

$$\mu_1 = \frac{0.21^2 \times 0.25^2}{3} = 9.2 \times 10^{-4}$$

Gate 2

$$P = P_{20} + P_{21}$$

$$\mu_2 = \frac{0.25}{2}(0.003 + 0.15) = 1.9 \times 10^{-2}$$

Gate 3

$$P = P_{10} P_{29}$$

$$\mu_3 = 0.01^2 \times 0.25^2 = 2 \times 10^{-6}$$

Gate 4

$$P = P_{10} P_{29} + P_{12}$$

$$\mu_4 = \mu_3 + 3.8 \times 10^{-4} = 3.8 \times 10^{-4}$$

Gate 5

$$P = (P_{10} P_{29} + P_{12}) P_5$$

$$\mu_5 = \mu_4 \times 1.3 \times 10^{-3} \times \frac{4}{3} = 6.6 \times 10^{-7}$$

Gate 6

$$P = (P_{10}P_{29} + P_{12})P_5 + P_7 + P_8$$

$$\mu_6 = \mu_5 + 3.8 \times 10^{-4} + 1.4 \times 10^{-2} = 1.44 \times 10^{-2}$$

Gate 7

$$P = (P_{20} + P_{21})(P_{10}P_{29} + P_{12})$$

$$\mu_7 = \mu_2\mu_6 \times \frac{4}{3} = 3.65 \times 10^{-4}$$

Gate 8

$$P = P_5 + P_7 + P_8$$

$$\mu_8 = 1.3 \times 10^{-3} + 3.8 \times 10^{-4} + 1.4 \times 10^{-2} = 1.57 \times 10^{-2}$$

Gate 9

$$P = P_{25} + P_{30}$$

$$\mu_9 = 2.6 \times 10^{-2} + 1.3 \times 10^{-3} = 2.73 \times 10^{-2}$$

Gate 10

$$P = (P_{25} + P_{30})P_{12}$$

$$\mu_{10} = \mu_9 \times 3.8 \times 10^{-4} \times \frac{4}{3} = 1.38 \times 10^{-5}$$

Gate 11

$$P = P_{29}P_{30}$$

$$\mu_{11} = 0.01^2 \times 0.25^2 = 2.01 \times 10^{-6}$$

Gate 12

$$P = P_{29}P_{30} + P_{25}$$

$$\mu_{12} = \mu_{11} + 2.6 \times 10^{-2} = 2.60 \times 10^{-2}$$

Gate 13

$$P = (P_{29}P_{30} + P_{25})P_{10}$$

$$\mu_{13} = \mu_{12} \times 1.3 \times 10^{-3} \times \frac{4}{3} = 4.50 \times 10^{-5}$$

Gate 14

$$P = (P_{25} + P_{30})P_{12}$$

$$\mu_{14} = \mu_{10} + \mu_{13}$$

$$= 1.38 \times 10^{-5} + 4.5 \times 10^{-5} = 5.88 \times 10^{-5}$$

Gate 15

$$P = (P_5 + P_7 + P_8)[(P_{25} + P_{30})P_{12} + (P_{29}P_{30} + P_{25})P_{10}]$$

$$\mu_{15} = \mu_8\mu_{14} \times \frac{4}{3}$$

$$= 1.5^7 \times 10^{-2} \times 5.88 \times 10^{-5} \times \frac{4}{3} = 1.23 \times 10^{-6}$$

Gate 16

$$P = P_{18}P_{30}$$

$$\mu_{16} = 1 \times 10^{-3} \times 1.3 \times 10^{-3} \times \frac{4}{3} = 1.70 \times 10^{-6}$$

Gate 17

$$P = P_{18}P_{30} + P_{20} + P_{21}$$

$$\mu_{17} = \mu_{16} + 3.8 \times 10^{-4} + 1.9 \times 10^{-2} = 1.94 \times 10^{-2}$$

Gate 18

$$P = P_2 + P_5$$

$$\mu_{18} = 2.6 \times 10^{-2} + 1.3 \times 10^{-3} = 2.73 \times 10^{-2}$$

Gate 19

$$P = (P_2 + P_5)P_{15}$$

$$\mu_{19} = \mu_{18} \times 1 \times 10^{-3} \times \frac{4}{3} = 3.60 \times 10^{-5}$$

Gate 20

$$P = (P_{20} + P_{21} + P_{18}P_{30})[(P_2 + P_5)P_{15}]$$

$$\mu_{20} = \mu_{17}\mu_{19} \times \frac{4}{3} = 1.94 \times 10^{-2} \times 3.6 \times 10^{-5} \times \frac{4}{3} = 1.00 \times 10^{-6}$$

Gate 21

$$P = P_{20} + P_{21} + P_{30}$$

$$\mu_{21} = 3.8 \times 10^{-4} + 1.9 \times 10^{-2} + 1.3 \times 10^{-3} = 2.07 \times 10^{-2}$$

Gate 22

$$P = P_{12} + P_{29}$$

$$\mu_{22} = 3.8 \times 10^{-4} + 1.3 \times 10^{-3} = 1.68 \times 10^{-3}$$

Gate 23

$$P = (P_{12} + P_{29})P_2$$

$$\mu_{23} = \mu_{22} \times 2.6 \times 10^{-2} \times \frac{4}{3} = 5.80 \times 10^{-5}$$

Gate 24

$$P = (P_{20} + P_{21} + P_{30})(P_{12} + P_{29})P_2$$

$$\mu_{24} = \mu_{21}\mu_{23} \times \frac{4}{3} = 2.07 \times 10^{-2} \times 5.8 \times 10^{-5}\frac{4}{3} = 1.60 \times 10^{-6}$$

Gate 25

$$P = P_7 + P_8$$

$$\mu_{25} = 3.8 \times 10^{-4} + 1.4 \times 10^{-2} = 1.40 \times 10^{-2}$$

Gate 26

$$P = (P_7 + P_8)P_{18}P_{30}$$

$$\mu_{26} = \mu_{25} \times 1 \times 10^{-3} \times 1.3 \times 10^{-3} \times 2 = (\ll 10^{-5})$$

Note that the factor 2 in the μ_{26} expression corrects the product of three FDTs to the resultant FDT.

Gate 27

$$P = P_5 P_{15}$$

$$\mu_{27} = 1.3 \times 10^{-3} \times 1 \times 10^{-3} \times \frac{4}{3} = 1.70 \times 10^{-6}$$

Gate 28

$$P = P_7 + P_8 + P_5 P_{15}$$

$$\mu_{28} = 3.8 \times 10^{-4} + 1.4 \times 10^{-2} + \mu_{27} = 1.44 \times 10^{-2}$$

Gate 29

$$P = P_{18} P_{25}$$

$$\mu_{29} = 1 \times 10^{-2} \times 2.6 \times 10^{-2} \times \frac{4}{3} = 3.50 \times 10^{-5}$$

Gate 30

$$P = (P_5 P_{15} + P_7 + P_8)(P_{18} P_{25})$$

$$\mu_{30} = \mu_{28} \mu_{29} \times \frac{4}{3} = 1.44 \times 10^{-2} \times 3.5 \times 10^{-5} \times \frac{4}{3} = (\ll 10^{-5})$$

Gate 31

$$\mu_0 = \mu_1 + \mu_7 + \mu_{15} + \mu_{20} + \mu_{24} + \mu_{26} + \mu_{30}$$

$$= (9.2 \times 10^{-4}) + (3.65 \times 10^{-4}) + (1.2 \times 10^{-6})$$

$$+ (1 \times 10^{-6}) + (1.6 \times 10^{-6}) + (\ll 10^{-5}) + (\ll 10^{-5})$$

$$\mu_0 = 1.28 \times 10^{-3}$$

C.9 COMMON-MODE FAILURE ANALYSIS

C.9.1 Terms of reference
- System schematic (Figure C.1);
- independent failure minimal cut-set listing (Table C.1);
- proof test interval $\tau = 0.25$ year;
- mathematical model (Figure C.3).

C.9.2 Common-mode analysis listing
The common-mode analysis listing is given in Table C.3. From the table, allowing a 25% time factor for concurrency of failures,

$$\text{Effective common-mode FDT} = 3.17 \times 10^{-3} \times 25\%$$
$$= 7.93 \times 10^{-4}$$

Table C.3 Common-mode analysis listing

Ident	Cut set	D_p (%)	β	Common-mode FDT μ_{CM}
1	P_2P_{25}	0	0.1	2.6×10^{-3}
2	$P_2P_{12}P_{20}$	33	0.02	7.6×10^{-6}
3	$P_2P_{12}P_{21}$	100	0.001	3.8×10^{-7}
4	$P_2P_{12}P_{30}$	83	0.002	8.4×10^{-7}
5	$P_2P_{20}P_{29}$	83	0.002	8.4×10^{-7}
6	$P_2P_{21}P_{29}$	83	0.002	2.6×10^{-6}
7	$P_2P_{29}P_{30}$	17	0.046	6.0×10^{-5}
8	$P_2P_{15}P_{20}$	92	0.001	3.8×10^{-7}
9	$P_2P_{15}P_{21}$	100	0.001	1.0×10^{-6}
10	$P_2P_{15}P_{18}P_{30}$	42	0.014	1.4×10^{-5}
11	$P_5P_{10}P_{25}$	17	0.046	6.0×10^{-5}
12	$P_5P_{10}P_{20}P_{29}$	17	0.046	1.7×10^{-5}
13	$P_5P_{10}P_{21}P_{29}$	17	0.046	6.0×10^{-5}
14	$P_5P_{10}P_{29}P_{30}$	0	0.1	1.3×10^{-4}
15	$P_7P_{10}P_{25}$	83	0.002	7.6×10^{-7}
16	$P_7P_{10}P_{29}P_{30}$	17	0.046	1.7×10^{-5}
17	$P_8P_{10}P_{25}$	83	0.002	2.6×10^{-6}
18	$P_8P_{10}P_{29}P_{30}$	17	0.046	6.0×10^{-5}
19	$P_5P_{12}P_{25}$	83	0.002	7.6×10^{-7}
20	$P_5P_{12}P_{20}$	33	0.02	7.6×10^{-6}
21	$P_5P_{12}P_{21}$	100	0.001	3.8×10^{-7}
22	$P_5P_{12}P_{30}$	33	0.02	7.6×10^{-6}
23	$P_7P_{12}P_{25}$	33	0.02	7.6×10^{-6}
24	$P_7P_{12}P_{30}$	33	0.02	7.6×10^{-6}
25	$P_8P_{12}P_{25}$	100	0.001	3.8×10^{-7}
26	$P_8P_{12}P_{30}$	100	0.001	3.8×10^{-7}
27	$P_7P_{18}P_{25}$	83	0.002	7.6×10^{-7}
28	$P_7P_{18}P_{30}$	83	0.002	7.6×10^{-7}
29	P_7P_{20}	0	0.1	3.8×10^{-5}
30	P_7P_{21}	100	0.001	3.8×10^{-7}
31	$P_8P_{18}P_{25}$	100	0.001	1.0×10^{-6}
32	$P_8P_{18}P_{30}$	100	0.001	1.0×10^{-6}
33	P_8P_{20}	100	0.001	3.8×10^{-7}
34	P_8P_{21}	100	0.001	1.4×10^{-5}
35	$P_5P_{15}P_{18}P_{25}$	83	0.002	2.0×10^{-6}
36	$P_5P_{15}P_{18}P_{30}$	17	0.046	4.6×10^{-5}
37	$P_5P_{15}P_{20}$	83	0.002	7.6×10^{-7}
38	$P_5P_{15}P_{21}$	100	0.001	1.0×10^{-6}
			Total	3.17×10^{-3}

C.9.3 Discussion arising from common-mode analysis

Table C.4 lists system common-mode failure FDTs μ_{CM} for a range of proof test intervals τ. The table shows that system common-mode failure cannot be improved to $<10^{-4}$ by more frequent proof testing in excess of three months.

Table C.4 Proof test interval/ system common-mode FDT

τ *(months)*	μ_{CM}
3	7.93×10^{-4}
2	5.28×10^{-4}

C.10 OVERALL SYSTEM ANALYSIS AND CONCLUSIONS

The mathematical model given in Figure C.3 derives an overall system failure probability on demand of 2.08×10^{-3} which is attributable to both independent and dependent failure modes. The system failure probability on demand of the assessed system does not meet the desired criterion of 10^{-4}.

C.10.1 Review of independent failure mode

- Failures due to independent mode yield a system FDT of 1.29×10^{-3} and accounts for 62% of the overall system FDT.
- The independent failure FDT μ_1 due to the process shutdown valves is assessed at 9.2×10^{-4} and accounts for 44% of the overall system FDT.
- The independent failure FDT μ_7 is assessed at 3.65×10^{-4} and accounts for 18% of the overall system FDT.
- The study shows that the level of redundancy in the process shutdown valve system is inadequate.
- The study shows that the most significant contributions to μ_7 are due to the single flowmeter channel F and the single temperature channel T. The level of redundancy in either F or T is inadequate.

C.10.2 Review of dependent failure mode

- Failures due to dependent mode yield a system FDT of 7.93×10^{-4} and accounts for 38% of the overall system FDT.

- The process shutdown valves provide, based on 25% concurrency of failures, a common-mode FDT of 6.5×10^{-4} which accounts for 82% of the system common-mode FDT and 31% of the overall system FDT.

C.10.3 Review of overall system

- Forty-four percent of the overall system FDT is attributable to independent failures related to the process shutdown valves.
- Eighteen percent of the overall system FDT is attributable to independent failures due to the flowmeter F and temperature sensor T.
- Thirty-eight percent of the overall system FDT is attributable to system common-mode failures μ_{CM}.

C.10.4 System assessment summary

For the system initially proposed the project aims as defined under section C.3 can now be stated.

- The system independent failure probability on demand has been assessed at 1.29×10^{-3}.
- The system dependent failure probability on demand has been assessed at 7.93×10^{-4}.
- The system overall failure probability on demand has been assessed at 2.08×10^{-3}.

C.11 OVERALL SYSTEM RECOMMENDATIONS

The following is recommended.

- The process shutdown valve system should be redesigned with improved redundancy and diversity in order to improve the subsystem independent and dependent failure contributions to the overall system FDT.
- Consideration should be given to a regime of staggered symmetrical proof testing in respect of the redesigned process valve shutdown system. This would yield a factor of three improvement in the common-mode contribution from this source (Table 5.3, three-out-of-three failure logic).
- Redundancy and diversity should be incorporated in the flowmeter channel F **or** temperature channel T primarily to improve the independent failure contributor μ_7 and hence the overall system independent FDT. Diversity is also recommended in order to offset degradation of β values in those cut sets where the flow parameter is present.

C.12 ACTIONS BASED ON RECOMMENDATIONS

C.12.1 Process shutdown valves

(a) Dependent failure mode

The first recommendation of section C.11 calls for improved redundancy and diversity in the process shutdown valve system. It is therefore logical to increase the number of shutdown valves from two to three in order to meet the independent failure requirement. This action alone, if the three valves remain identical, will not improve their common-mode FDT contribution which would remain at $\mu = 6.5 \times 10^{-4}$ in 25% concurrency of failure terms. If however, the shutdown valves were mutually diverse, an improved β factor corresponding to the degree of partial diversity would be attainable.

Hence for the proposed three-valve system and in order to realize adequate diversity it is expedient to examine a minimal cut set which is representative of the proposal. Let the cut set be conveniently defined as '*ABC*' respectively. In assessing mutual diversity between the three valves, account will need to be taken of commonalities which may be present in manufacture, design, environment, maintenance and proof testing.

Diversities are now assessed in the following areas.

Manufacture
The three valves would be supplied by three independent manufacturers. It is assumed that mutual diversity between types could achieve a figure of 90% which recognizes the possibility of some special bought out common items in the three types.

Design
Design of the valves would differ as far as possible. For example, types of actuator could be say diaphragm, cylinder, rotary action etc.

The valves themselves could be globe, ball or diaphragm type with the necessary quick closure characteristics. It is assumed that 50% mutual diversity could be achieved between the three types.

Process environment
The valves occupy the same point in the process stream and hence all wetted parts must be identical in material terms. Significant diversity is not therefore seen to be attainable in this respect.

Maintenance
This facet is under the direct control of the user and hence each valve could be maintained by different personnel. In terms of maintenance it

would be possible to achieve a mutual diversity between the three types of 75%. This acknowledges that commonalities are likely to exist in workshop and management practices.

Proof testing
Staggered proof testing carried out by different personnel would contribute significantly to diversity. Also although a common philosophy of testing is likely to be applicable to each valve there could be variations in the actual tests due to differences in valve manufacture, e.g. timing of closures, method of application of operating media, verification of closure etc., all of which would further enhance diversity. With the above points taken into consideration a 75% mutual diversity between types could be achievable.

Mutual diversity
The average of the above would indicate the level of mutual diversity across the three types which could be reasonably expected.

Manufacture	90%
Equipment design	50%
Process environment	0%
Maintenance	75%
Proof testing	75%
Average	58%

The above assesses the single-group common-mode failure of the proposed three-valve system represented by the cut set at ident 1 in the listing in Table C.3 noting that the cut set has an additional member element due to the proposed third shutdown valve. Automatic valve shutdown systems are inherently significant contributors to safety system failure probabilities and in consequence a more detailed assessment of common mode should take into account dependent failures attributable to partial common-mode set groups in accordance with the following. The analyses will be based on the mutual type diversity evaluated above.

Initially the common-mode contribution from the revised cut set at ident 1 of Table C.3 will be evaluated to give the dependent failure of the three-valve group.

Referring to section 10.7 the group common-mode dependent failure arising from the parent cut set ABC is evaluated as follows.

$$S_p = \text{No. of element pairs } ABC = \frac{3(3-1)}{2} = 3$$

Active element cut set $= ABC$ (all elements mutually diverse)

Therefore

$$S_a = \text{No. of active element pairs } = \frac{3(3-1)}{2} = 3$$

From the above analysis, mutual diversity between classes is 58%.

$$d = \text{Degrees of diversity: 3 active pairs} = 0.58 + 0.58 + 0.58$$
$$= 1.74$$

Percentage diversity D_p across cut set ABC:

$$D_p = \frac{d}{S_p} \times 100\% = \frac{1.74 \times 100}{3} = 58\%$$

Therefore

$$\beta \text{ factor} = 0.007$$

Valve system common-mode FDT:

$$\mu_{CM} = |ABC| + A|BC| + B|AC| + C|AB|$$
$$= [(\text{Group dep. failure}) + (\text{Ind. } A \times \text{Dep. } BC) + (\text{Ind. } B \times \text{Dep. } AC)$$
$$+ (\text{Ind. } C \times \text{Dep. } AB)] \times (\text{Concurrency factor } 0.25)$$
$$= \{(1.82 \times 10^{-4}) + \tfrac{4}{3}[(2.6 \times 10^{-2})(2.6 \times 10^{-2} \times 0.007)]3\}0.25$$
$$= 5.02 \times 10^{-5}$$

Note that the factor $\tfrac{4}{3}$ corrects the direct product of two FDTs, i.e. in general terms

$$\mu_0 = \frac{4}{3}\mu_1\mu_2$$

The original non-diverse shutdown system consisting of two valves provided a system common-mode FDT contribution of 6.5×10^{-4}. Modifying the valve system by the addition of a third valve accompanied by the establishment of 58% mutual diversity between valves would result in an improved valve system common-mode FDT. Based on the specified proof test interval of three months, the proposed shutdown system would contribute 5.02×10^{-5} to the overall system common-mode failure. This would improve the overall system dependent failure FDT from 7.93×10^{-4} to 1.93×10^{-4}.

(b) Independent failure mode

In terms of independent failure the three-valve system will yield an improved FDT, namely

$$\frac{0.21^3 \times \tau^3}{4} = 3.6 \times 10^{-5}$$

C.12.2 Flowmeter channel F

(a) Independent failure mode

In accordance with the third recommendation of section C.11, duplication of the flowmeter channel F to provide single redundancy with diversity would improve the independent failure FDT given as μ_7 in the mathematical model. Referring to the model in Figure C.3 the output μ_6 from gate 6 would be revised as follows:

$$\mu_6 = (\text{Output from gate 5}) + P_7 + (P_8)^2$$

$$= (<10^{-5}) + (3.8 \times 10^{-4}) + \frac{4}{3}(1.4 \times 10^{-2})^2$$

$$= 6.4 \times 10^{-4}$$

Hence output from gate 7 is

$$\mu_7 = (\text{Output from gate 6}) \times (\text{Output from gate 2}) \times \frac{4}{3}$$

$$= \mu_6 \times \mu_2 \times \frac{4}{3}$$

$$= (6.4 \times 10^{-4})(1.9 \times 10^{-2}) \times \frac{4}{3}$$

$$= 1.62 \times 10^{-5}$$

(b) Dependent failure mode

The common-mode failure analysis listing shows the presence of flowmeter F as P_8 in cut sets 17, 18, 31, 32, 33 and 34. These cut sets, in accordance with the third recommendation of section C.11 would each contain two redundant flowmeter elements. Reassessment of the cut sets shows that if a 50% diversity can be achieved between the redundant flowmeters, then the common-mode contribution from the stated sets to the overall system common-mode failure will remain virtually unchanged at 1.35×10^{-5}. Failure to provide 50% diversity will worsen the overall system common-mode FDT by an amount equal to 1.09×10^{-4}.

C.12.3 System assessment summary review

For the system initially proposed the project aims as defined under section C.3 can now be stated.

- The system independent failure probability on demand has been assessed at 1.29×10^{-3}.
- The system dependent failure probability on demand has been assessed at 7.93×10^{-4}.
- The system overall failure probability on demand has been assessed at 2.08×10^{-3}.
- The system as designed does not meet the desired criterion of 10^{-4}.

C.13 REASSESSMENT BASED ON RECOMMENDATIONS

Implementation of recommendations in section C.11 would yield the following improvements in safety system failure probability on demand.

- The independent failure FDT is reassessed at 5.74×10^{-5}.
- The dependent failure FDT is reassessed at 1.93×10^{-4}.
- The system overall FDT is reassessed at 2.50×10^{-4}.

C.14 DISCUSSION

The assessment indicates that a significant limiting factor in the achievement of an overall system 10^{-4} FDT criterion is attributable to common-mode failure. It has been shown that if the recommendations are implemented, then the safety system FDT does not meet the 10^{-4} criterion. There are thus two initial options for consideration, as follows.

- Proof test the complete safety system at six-week intervals instead of three months. This would produce an overall system FDT of 1.16×10^{-4} which would satisfy the criterion.
- Proof test the shutdown system only at six-week intervals under a regime of staggered symmetrical testing. This would produce a safety system common-mode FDT of 1.5×10^{-4} and hence an overall system FDT of 2.07×10^{-4}.

C.15 FINAL COMMENTS

Implementation of the first option would satisfy the desired safety criterion but would entail higher charges on plant maintenance. The achievement of a safety system FDT criterion of 10^{-4} is synonymous with higher standards of safety engineering design and maintenance. Any further improvement would be dependent on the nature of the plant hazard, the process demand rate, and the degree of risk which would be acceptable in the particular safety case. Generally a safety system FDT criterion necessitates compatibility with the process demand rate so as to realize a given risk criterion to meet a particular hazard.

In major hazard scenarios risk criteria are mainly met by optimizing demand rate with safety system FDT. There is thus a further option whereby the overall process demand rate could be assessed with a view to reducing its incidence. This measure frequently involves penalties in the nature of higher costs related to plant equipment when compared with extended safety system capital and running costs.

The study demonstrates the effectiveness of a powerful common-mode analysis technique through minimal cut sets generated from fault trees and identifies precisely where defences need to be located. This has obvious advantages in respect of monetary savings in both capital and maintenance costs over those methods which rely on overdesign of defences with little knowledge of their most effective locations.

D

Reliability case study of an automatic fire valve based on failure mode and effect analysis

D.1 INTRODUCTION

A unit valve system assembly is proposed as a fire protection device for which field data as a composite unit, due to its novel design, are not available. As a necessary adjunct to assessing its suitability in reliability terms, a failure mode and effect analysis is used to determine its composite dangerous and safe failure rates.

D.2 PROJECT DESCRIPTION

The automatic fire protection valve is shown in Figure D.1 which uses a metal alloy thermal trigger, actuated by a predetermined higher ambient temperature resulting from a conflagration in the local process environment. Such a device, to be reliable, must operate from a guaranteed power source, be adequately responsive to demands and maintain the closure state for an adequate period commensurate with a criterial safety requirement. Figure D.1 shows a normally open valve having a valve seat with two machined recesses arranged circumferentially on the inner face. The purpose of the recesses is to receive a flexible circular diaphragm attached to the top of the valve plug. When the valve operates, the plug enters the seat to provide closure, and at the same time the flexible diaphragm will spring into the lower recess and hence lock the plug into the seat. By this means, valve closure will no longer be reliant on the actuating spring, which is expected to lose its spring strength in consequence of high temperature conditions caused by the surrounding fire. In the event of any foreign obstruction in the open

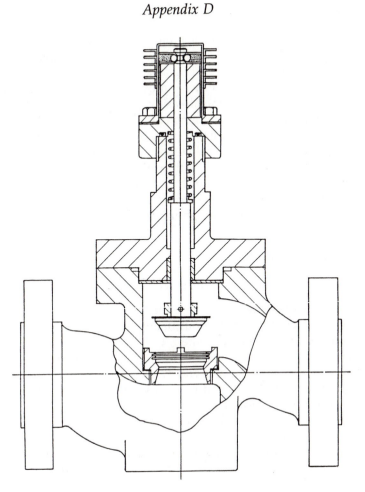

Figure D.1 Automatic fire protection valve.

seat, the diaphragm would be able to enter the upper recess and again lock the plug into the seat.

The upper assembly contains valve plug, stem, operating spring and thermal trigger and is bolted as a primed unit onto a standard valve body. The subassembly, which houses the thermal trigger, is likewise bolted on to the top of the upper assembly such that in the primed state the operating spring is held compressed against the restraining thermal trigger. The trigger itself is a metal alloy which is arranged to plasticize at a given ambient temperature with very close tolerances over a wide range from 30 °C. At the required ambient temperature, under the pressure of the operating spring, the metal balls move outwards from the upper valve circumferential recess into the now plastic trigger, thus allowing downward travel of the valve plug assembly into the seat and hence isolating the hazardous process medium upstream of the valve. In order to improve dynamic response, the trigger assembly has an outer metal casing with fins to promote a rapid heat flow from the surrounding ambient. The heat flow is directed preferentially towards the thermal trigger by the addition of a thermal insulator placed immediately below and around the valve spindle. The valve is regarded as a sacrificial item in that its only function is to isolate dangerous materials in major fire incidents for a time long enough for site evacuation to be completed and also to allow emergency services to take possible action in fighting the fire.

D.3 PROJECT AIMS

The unit valve system assessment is required to provide:

- the unit system dangerous failure rate;
- the unit system safe, or spurious, failure rate;
- the unit system total failure rate;
- system mean fractional dead times due to dangerous failures for proof test intervals of 1, 3, 6 and 12 months;
- system mean fractional dead times resulting from times to carry out scheduled proof tests, namely 0.5, 1, 2 and 5 hours;
- system mean fractional dead times due to mean repair times of 5, 12, 24 and 48 hours;
- compilation of overall mean fractional dead times for the stated range of proof test intervals, proof testing times and a mean repair time of 48 hours;
- the frequency of safe spurious operations;
- proposition of a valid proof test procedure;
- conclusions.

D.4 LIST OF SYMBOLS

θ_D = unrevealed dangerous failure rate of the valve unit

θ_R = revealed safe failure rate of the valve unit

θ_0 = overall failure rate of the valve unit

τ = proof test interval

τ_r = mean repair time

τ_t = mean time of proof test

D.5 ASSESSMENT TERMS OF REFERENCE

In all evaluations:

- failure rates are in terms of faults per annum;
- proof test intervals are in terms of years;
- times to carry out proof tests are in terms of years.

D.6 ASSESSMENT OF UNIT SYSTEM DANGEROUS FAILURE RATE

The valve system is seen to contain three principal components which could, because of unrevealed failures prevent the valve from operating when a demand (fire incident) occurs. These items are considered as follows, using typical data shown in Table D.1.

- Operating spring: this fails due to

  (a) loss of spring tension;
  (b) fracture.

  These two modes would be expected to yield a typical failure rate of 5×10^{-4} faults per annum.
- Gland stiction: a gland which is not protected by a bellows seal would be expected to yield a typical failure rate of 5×10^{-3} faults per annum.
- Bottom guide: a bottom guide designed with adequate clearance and with designed reliance on self-centring of the plug as it enters the valve seat would be expected to yield a typical failure rate of 2.5×10^{-3} faults per annum.

The overall dangerous unrevealed failure rate is obtained from the sum of the above component rates which equates to 8.0×10^{-3} faults per annum.

Table D.1 Component data base

Component	Failure mode	Failure rate per 10^6 hours	Failure rate per annum
Spring	Loss of tension and fracture	0.06	5×10^{-4}
Gland	Stiction	0.57	5×10^{-3}
Bottom guide	Misalignment	0.29	2.5×10^{-3}

D.7 ASSESSMENT OF UNIT SYSTEM SAFE FAILURE RATE

These failures are revealed in that they initiate valve system operation in the absence of a fire incident demand. The valve model described contains two principal elements by which spurious operation would occur:

- creep of the thermal trigger;
- fracture of the valve spindle at the retaining balls.

It is considered that the former failure mode would be the more significant and that this would occur once per year out of a population of 2000 valves. This is based on observations carried out over a period of five years on two model valves which have experienced wide variations in normal ambient temperatures without any detectable creep. The spurious failure rate is therefore assessed at 5.0×10^{-4} faults per annum.

D.8 ASSESSMENT OF UNIT SYSTEM OVERALL FAILURE RATE

The overall failure rate of the valve comprises the unrevealed dangerous and revealed safe failure rates and is of significance in assessing the system mean fractional dead time contribution arising from repair times. During repair, the local environment would be without fire protection, and hence repair time will need to be taken into account in assessing the overall dangerous fractional dead time. Repair regimes are invoked when faults are revealed by proof tests or from spurious operations and will always result in repair downtimes.

On the basis of the above assessed figures for the unrevealed and revealed failure rates the overall valve failure rate is assessed at 8.5×10^{-3} faults per annum.

D.9 MEAN FRACTIONAL DEAD TIMES DUE TO DANGEROUS FAILURES

FDT of valve unit system

$$= \frac{\text{(Unrevealed dangerous failure rate)} \times \text{(Proof test interval)}}{2}$$

$$= \frac{\theta_D \tau}{2}$$

Table D.2 gives mean fractional dead times (FDTs) for the stated range of proof test intervals.

Table D.2 System FDTs/proof test
intervals

Proof test interval (months)	System FDT
1	3.3×10^{-4}
3	1.0×10^{-3}
6	2.0×10^{-3}
12	4.0×10^{-3}

D.10 MEAN FRACTIONAL DEAD TIMES DUE TO PROOF TEST TIMES

During proof testing the local environment is without protection should a fire incident occur. The test regime is therefore seen to be that of on-line testing as previously discussed in section 4.4.1. The duration of the test represents a statistical downtime and therefore is a component of the overall fractional dead time of the valve.

$$\text{Testing time FDT} = \frac{\text{Test duration}}{\text{Proof test interval}}$$

$$= \frac{\tau_t}{\tau}$$

Table D.3 gives FDT values for the stated ranges of proof test intervals and test times.

Table D.3 System FDTs – test times/proof test intervals

Test time (hours)	Proof test intervals (months)			
	1	3	6	12
0.5	6.85×10^{-4}	2.28×10^{-4}	1.14×10^{-4}	5.70×10^{-5}
1.0	1.37×10^{-3}	4.57×10^{-4}	2.28×10^{-4}	1.14×10^{-4}
2.0	2.74×10^{-3}	9.13×10^{-4}	4.57×10^{-4}	2.28×10^{-4}
5.0	6.85×10^{-3}	2.28×10^{-3}	1.14×10^{-3}	5.71×10^{-4}

D.11 MEAN FRACTIONAL DEAD TIMES DUE TO REPAIR TIMES

All failures, whether unrevealed or revealed will necessitate repair. During repair downtime the protective system is unavailable for fire protection.

Repair time FDT = (System overall failure rate) × (Mean repair time)

$$= \theta_0 \tau_r$$

Table D.4 gives FDT values for the stated range of mean repair times.

Table D.4 Mean repair times/system FDTs

Mean repair time (hours)	FDT
5	4.85×10^{-6}
12	1.16×10^{-5}
24	2.32×10^{-5}
48	4.66×10^{-5}

D.12 SYSTEM OVERALL MEAN FRACTIONAL DEAD TIMES

An overall system mean fractional dead time is given by the FDTs due to:

- valve dangerous failures;
- proof testing time;
- repair time.

Table D.5 gives system FDTs for the stated range of proof test intervals, test times and a mean repair time of 48 hours.

D.13 FREQUENCY OF SAFE (SPURIOUS) OPERATIONS

The assessment in section D.7 has concluded a spurious failure rate of 5×10^{-4} faults per annum. The reciprocal of this figure gives a statistical period of 2000 years. This is interpreted on average, as one such operation in 2000 years for the given valve or one spurious operation per year out of a population of 2000 units which would be identical in terms of design, construction, maintenance, testing and duty environment.

Table D.5 System FDTs – proof test intervals/test times

Proof test interval (months)	Test time (hours)	FDTs			
		Valve	Test	Repair	Overall
1	0.5	3×10^{-4}	6.9×10^{-4}	4.7×10^{-5}	1.0×10^{-3}
1	1.0	3×10^{-4}	1.4×10^{-3}	4.7×10^{-5}	1.7×10^{-3}
1	2.0	3×10^{-4}	2.7×10^{-3}	4.7×10^{-5}	3.1×10^{-3}
1	5.0	3×10^{-4}	6.9×10^{-3}	4.7×10^{-5}	7.2×10^{-3}
3	0.5	1×10^{-3}	2.3×10^{-4}	4.7×10^{-5}	1.3×10^{-3}
3	1.0	1×10^{-3}	4.6×10^{-4}	4.7×10^{-5}	1.5×10^{-3}
3	2.0	1×10^{-3}	9.1×10^{-4}	4.7×10^{-5}	2.0×10^{-3}
3	5.0	1×10^{-3}	2.3×10^{-3}	4.7×10^{-5}	3.3×10^{-3}
6	0.5	2×10^{-3}	1.1×10^{-4}	4.7×10^{-5}	2.2×10^{-3}
6	1.0	2×10^{-3}	2.3×10^{-4}	4.7×10^{-5}	2.3×10^{-3}
6	2.0	2×10^{-3}	4.6×10^{-4}	4.7×10^{-5}	2.5×10^{-3}
6	5.0	2×10^{-3}	1.1×10^{-3}	4.7×10^{-5}	3.2×10^{-3}
12	0.5	4×10^{-3}	5.7×10^{-5}	4.7×10^{-5}	4.1×10^{-3}
12	1.0	4×10^{-3}	1.1×10^{-4}	4.7×10^{-5}	4.2×10^{-3}
12	2.0	4×10^{-3}	2.3×10^{-4}	4.7×10^{-5}	4.3×10^{-3}
12	5.0	4×10^{-3}	5.7×10^{-4}	4.7×10^{-5}	4.6×10^{-3}

D.14 PROPOSALS FOR A PROOF TEST PROCEDURE

An ideal proof test of the valve system would be satisfied by generating a real demand under the most severe environmental dynamic conditions and thence to observe that successful process line closure had taken place within the designed performance boundary of the valve unit.

Clearly it is not practicable to initiate a real demand and hence simulation must be employed in order to carry out a representative proof test. The valve system under consideration would be expected to have a service lifetime of, say 20 years and if such a long interval were to be regarded as a test interval in order to dispense with proof testing, then the failure probability on demand would be 7.6×10^{-2} which, in most cases, would be unacceptable. For the majority of situations, subject to the nature of a given hazard and in the absence of demand assessment, an acceptable fractional dead time would need to meet a value of at least 10^{-3} or better and therefore representative proof testing would be mandatory.

A practical proposal for a proof test procedure would be to separately test the topwork thermal trigger by removing the assembly from the valve body proper and attaching it to a suitable test rig. On the rig, the thermal trigger could be subjected to a representative temperature ramp and its action and time response observed. This would test the operating spring, gland and bottom guide and also verify that the thermal trigger assembly conformed to specification. After test, the thermal trigger would be replaced by a replacement trigger taken from the same batch of alloy as the original one. Also as part of the proof test, the valve seat would also undergo visual inspection to ensure absence of foreign materials which may obstruct closure. In order to carry out the test in minimum time it should be possible to substitute the trigger assembly to be tested with a previously tested unit. This would contribute to a lower test time FDT component of the overall system FDT.

A final important point would be that the valve would need to be isolated and bypassed in the process line, or the line isolated and drained downstream of the unit before removing its topwork. For maximum convenience the proof test could be carried out at annual process plant shutdown, which would represent an off-line testing regime with consequential system benefits due to elimination of the FDT test time component and cost savings in maintenance due to lower frequency testing.

D.15 CONCLUSIONS

- Table D.5 shows that overall mean fractional dead times from 1×10^{-3} to 4.6×10^{-3} can be achieved over wide ranges of proof test intervals and test durations and a common mean repair time of 48 hours.
- For proof test intervals in excess of one month, the system FDT is largely independent of the test and mean repair times.
- The range of assessed system FDTs indicates that the valve is capable, in reliability terms, of meeting the requirements of reliability criteria from moderate to more major hazard scenarios, subject to frequency of demands.
- The spurious operation of the valve is assessed to occur once in 2000 years or conversely, on average, there would be one spurious operation per year out of a population of 2000 valve units.
- If a system FDT worse than 10^{-3} could be acceptable, subject to a favourable demand rate, then a proof test at an annual plant shutdown would represent cost savings in those maintenance charges associated with testing.

Index